Teaching Tec

How to Make Lesso
and Build a Teaching Community
around Them

Greg Wilson

CRC Press
Taylor & Francis Group
Boca Raton London New York

CRC Press is an imprint of the
Taylor & Francis Group, an **informa** business

A CHAPMAN & HALL BOOK

CRC Press
Taylor & Francis Group
6000 Broken Sound Parkway NW, Suite 300
Boca Raton, FL 33487-2742

Printed and bound by CPI Group (UK) Ltd, Croydon, CR0 4YY on acid-free paper

International Standard Book Number-13: 978-0-367-35297-4 (Paperback)
                                      978-0-367-35328-5 (Hardback)

Library of Congress Cataloging-in-Publication Data

Names: Wilson, Greg, 1963- author.
Title: Teaching tech together : how to make lessons that work and build a teaching community around them / by Greg Wilson.
Description: Boca Raton : CRC Press, [2020] | Includes bibliographical references and index. | Summary: "Hundreds of grassroots groups have sprung up around the world to teach programming, web design, robotics, and other skills outside traditional classrooms. These groups exist so that people don't have to learn these things on their own, but ironically, their founders and instructors are often teaching themselves how to teach. There's a better way. This book presents evidence-based practices that will help you create and deliver lessons that work and build a teaching community around them. Topics include the differences between different kinds of learners, diagnosing and correcting misunderstandings, teaching as a performance art, what motivates and demotivates adult learners, how to be a good ally, fostering a healthy community, getting the word out, and building alliances with like-minded groups. The book includes over a hundred exercises that can be done individually or in groups, over 350 references, and a glossary to help you navigate educational jargon"– Provided by publisher.
Identifiers: LCCN 2019030229 (print) | LCCN 2019030230 (ebook) | ISBN 9780367352974 (paperback) | ISBN 9780367353285 (hardback) | ISBN 9780429330704 (ebook)
Subjects: LCSH: Computer programming–Study and teaching. | Web sites–Design–Study and teaching. | Robotics–Study and teaching.
Classification: LCC QA76.6 .W56 2020 (print) | LCC QA76.6 (ebook) | DDC 005.1–dc23
LC record available at https://lccn.loc.gov/2019030229
LC ebook record available at https://lccn.loc.gov/2019030230

Visit the Taylor & Francis Web site at
http://www.taylorandfrancis.com

and the CRC Press Web site at
http://www.crcpress.com

Teaching Tech Together

How to Make Lessons That Work and Build a Teaching Community around Them

Dedication

For my mother, Doris Wilson,
who taught hundreds of children to read and to believe in themselves.

And for my brother Jeff, who did not live to see it finished.
"Remember, you still have a lot of good times in front of you."

All royalties from the sale of this book are being donated to
the Carpentries,
a volunteer organization that teaches
foundational coding and data science skills
to researchers worldwide.

Contents

The Rules

1. Be kind: all else are details.

2. Remember that you are not your learners...

3. ...most people would rather fail than change...

4. ...ninety percent of magic consists of knowing one extra thing.

5. Never teach alone.

6. Never hesitate to sacrifice truth for clarity.

7. Make every mistake a lesson.

8. Remember that no lesson survives first contact with learners...

9. ...every lesson is too short for the teacher and too long for the learner...

10. ...nobody will be more excited about the lesson than you are.

1 Introduction

Grassroots groups have sprung up around the world to teach programming, web design, robotics and other skills to **free-range learners**. These groups exist so that people don't have to learn these things on their own, but ironically, their founders and teachers are often teaching themselves how to teach.

There's a better way. Just as knowing a few basic facts about germs and nutrition can help you stay healthy, knowing a few things about cognitive psychology, instructional design, inclusivity and community organization can help you be a more effective teacher. This book presents key ideas you can use right now, explains why we believe they are true and points you at other resources that will help you go further.

Re-Use

Parts of this book were originally created for the Software Carpentry instructor training program[1], and all of it can be freely distributed and re-used under the Creative Commons Attribution-NonCommercial 4.0 license (Appendix A). You can use the online version at http://teachtogether.tech/ in any class (free or paid) and can quote short excerpts under fair use[2] provisions, but cannot republish large parts in commercial works without prior permission.

Contributions, corrections, and suggestions are all welcome, and all contributors will be acknowledged each time a new version is published. Please see Appendix C for details and Appendix B for our code of conduct.

1.1 WHO YOU ARE

Section 6.1 explains how to figure out who your learners are. The four that this book is for all **end-user teachers**: teaching isn't their primary occupation, they have little or no background in pedagogy and they may work outside institutional classrooms.

Emily trained as a librarian and now works as a web designer and project manager in a small consulting company. In her spare time she helps run web design classes for women entering tech as a second career. She is now recruiting colleagues to run more classes in her area, and wants to know how to make lessons others can use and grow a volunteer teaching organization.

Moshe is a professional programmer with two teenage children whose school doesn't offer programming classes. He has volunteered to run a monthly after-school programming club, and while he frequently gives presentations to col-

---

[1] http://carpentries.github.io/instructor-training/

[2] https://en.wikipedia.org/wiki/Fair_use

leagues, he has no classroom experience. He wants to learn how to build effective lessons in reasonable time, and would like to know more about the pros and cons of self-paced online classes.

Samira is an undergraduate in robotics who is thinking about becoming a teacher after she graduates. She wants to help at weekend robotics workshops for her peers, but has never taught a class before and feels a lot of **impostor syndrome**. She wants to learn more about education in general in order to decide if it's for her, and is also looking for specific tips to help her deliver lessons more effectively.

Gene is a professor of computer science. They have been teaching undergraduate courses on operating systems for six years, and increasingly believe that there has to be a better way. The only training available through their university's teaching and learning center is about posting assignments and submitting grades in the online learning management system, so they want to find out what else they should be asking for.

These people have *a variety of technical backgrounds* and *some previous teaching experience*, but *no formal training in teaching, lesson design or community organization*. Most work with *free-range learners* and are *focused on teenagers and adults* rather than children; all *have limited time and resources*. We expect our quartet to use this material as follows:

Emily will take part in a weekly online reading group with her volunteers.

Moshe will cover part of this book in a one-day weekend workshop and study the rest on his own.

Samira will use this book in a one-semester undergraduate course with assignments, a project and a final exam.

Gene will read the book on their own in their office or while commuting, wishing all the while that universities did more to support high-quality teaching.

1.2   WHAT TO READ INSTEAD

If you are in a hurry or want a taste of what this book will cover, [Brow2018] presents ten evidence-based tips for teaching computing. You may also enjoy:

- The Carpentries instructor training[3], from which this book is derived.
- [Lang2016] and [Hust2012], which are short and approachable, and which connect things you can do right now to the research that backs them.
- [Berg2012; Lemo2014; Majo2015; Broo2016; Rice2018; Wein2018b] are all full of practical suggestions for things you can do in your classroom, but may make more sense once you have a framework for understanding why their ideas work.
- [DeBr2015], which explains what's true about education by explaining what isn't, and [Dida2016], which grounds learning theory in cognitive psychology.

---

[3]http://carpentries.github.io/instructor-training/

- [Pape1993], which remains an inspiring vision of how computers could change education. Andy Ko's excellent description[4] does a better job of summarizing Papert's ideas than I possibly could, and [Craw2010] is a thought-provoking companion to both.
- [Gree2014; McMi2017; Watt2014] explain why so many attempts at educational reform have failed over the past forty years, how for-profit colleges are exploiting and exacerbating the growing inequality in our society, and how technology has repeatedly failed to revolutionize education.
- [Brow2007] and [Mann2015], because you can't teach well without changing the system in which we teach, and you can't do that on your own.

Those who want more academic material may also find [Guzd2015a; Hazz2014; Sent2018; Finc2019; Hpl2018] rewarding, while Mark Guzdial's blog[5] has consistently been informative and thought-provoking.

1.3 ACKNOWLEDGMENTS

This book would not exist without the contributions of Laura Acion, Jorge Aranda, Mara Averick, Erin Becker, Azalee Bostroem, Hugo Bowne-Anderson, Neil Brown, Gerard Capes, Francis Castro, Daniel Chen, Dav Clark, Warren Code, Ben Cotton, Richie Cotton, Karen Cranston, Katie Cunningham, Natasha Danas, Matt Davis, Neal Davis, Mark Degani, Tim Dennis, Michael Deutsch, Brian Dillingham, Kathi Fisler, Denae Ford, Auriel Fournier, Bob Freeman, Nathan Garrett, Mark Guzdial, Rayna Harris, Ahmed Hasan, Ian Hawke, Felienne Hermans, Kate Hertweck, Toby Hodges, Mike Hoye, Dan Katz, Christina Koch, Shriram Krishnamurthi, Katrin Leinweber, Colleen Lewis, Lenny Markus, Sue McClatchy, Jessica McKellar, Ian Milligan, Julie Moronuki, Lex Nederbragt, Aleksandra Nenadic, Jeramia Ory, Joel Ostblom, Elizabeth Patitsas, Aleksandra Pawlik, Sorawee Porncharoenwase, Emily Porta, Alex Pounds, Thomas Price, Danielle Quinn, Ian Ragsdale, Erin Robinson, Rosario Robinson, Ariel Rokem, Pat Schloss, Malvika Sharan, Florian Shkurti, Dan Sholler, Juha Sorva, Igor Steinmacher, Tracy Teal, Tiffany Timbers, Richard Tomsett, Preston Tunnell Wilson, Matt Turk, Fiona Tweedie, Anelda van der Walt, Stéfan van der Walt, Allegra Via, Petr Viktorin, Belinda Weaver, Hadley Wickham, Jason Williams, Simon Willison, Karen Word, John Wrenn, and Andromeda Yelton. I am also grateful to Lukas Blakk for the logo, to Shashi Kumar for LaTeX help, to Markku Rontu for making the diagrams look better, and to everyone who has used this material over the years. Any mistakes that remain are mine.

---

[4]https://medium.com/bits-and-behavior/mindstorms-what-did-papert-argue-and-what-does-it-mean-for-learning-and-education-c8324b58aca4

[5]http://computinged.wordpress.com

1.4 EXERCISES

Each chapter ends with a variety of exercises that include a suggested format and how long they usually take to do in person. Most can be used in other formats—in particular, if you are going through this book on your own, you can still do many of the exercises that are intended for groups—and you can always spend more time on them than what's suggested.

If you are using this material in a teacher training workshop, you can give the exercises below to participants a day or two in advance to get an idea of who they are and how best you can help them. Please read the caveats in Section 9.4 before doing this.

HIGHS AND LOWS (WHOLE CLASS/5)

Write brief answers to the following questions and share with your peers. (If you are taking notes together online as described in Section 9.7, put your answers there.)

1. What is the best class or workshop you ever took? What made it so good?
2. What was the worst one? What made it so bad?

KNOW THYSELF (WHOLE CLASS/10)

Share brief answers to the following questions with your peers. Record your answers so that you can refer back to them as you go through the rest of this book.

1. What do you most want to teach?
2. Who do you most want to teach?
3. Why do you want to teach?
4. How will you know if you're teaching well?
5. What do you most want to learn about teaching and learning?
6. What is one specific thing you believe is true about teaching and learning?

WHY LEARN TO PROGRAM? (INDIVIDUAL/20)

Politicians, business leaders and educators often say that people should learn to program because the jobs of the future will require it. However, as Benjamin Doxtdator pointed out[6], many of those claims are built on shaky ground. Even if they were true, education shouldn't prepare people for the jobs of the future: it should give them the power to decide what kinds of jobs there are and to ensure that those jobs are worth doing. And as Mark Guzdial points out[7], there are actually many reasons to learn how to program:

---

[6]http://www.longviewoneducation.org/field-guide-jobs-dont-exist-yet/

[7]https://computinged.wordpress.com/2017/10/18/why-should-we-teach-programming-hint-its-not-to-learn-problem-solving/

1. To understand our world.
2. To study and understand processes.
3. To be able to ask questions about the influences on their lives.
4. To use an important new form of literacy.
5. To have a new way to learn art, music, science and mathematics.
6. As a job skill.
7. To use computers better.
8. As a medium in which to learn problem-solving.

Draw a 3×3 grid whose axes are labelled "low," "medium" and "high" and place each reason in one sector according to how important it is to you (the X axis) and to the people you plan to teach (the Y axis).

1. Which points are closely aligned in importance (i.e., on the diagonal in your grid)?
2. Which points are misaligned (i.e., in the off-diagonal corners)?
3. How should this affect what you teach?

2 Mental Models and Formative Assessment

The first task in teaching is to figure out who your learners are. Our approach is based on the work of researchers like Patricia Benner, who studied how nurses progress from novice to expert [Benn2000]. Benner identified five stages of cognitive development that most people go through in a fairly consistent way. For our purposes, we will simplify this progression to three stages:

Novices don't know what they don't know, i.e., they don't yet have a usable mental model of the problem domain.

Competent practitioners have a mental model that's adequate for everyday purposes. They can do normal tasks with normal effort under normal circumstances and have some understanding of the limits to their knowledge (i.e., they know what they don't know).

Experts have mental models that include exceptions and special cases, which allows allows them to handle situations that are out of the ordinary. We will discuss expertise in more detail in Chapter 3.

So what *is* a **mental model**? As the name suggests, it is a simplified representation of the most important parts of some problem domain that is good enough to enable problem solving. One example is the ball-and-spring models of molecules used in high school chemistry. Atoms aren't actually balls, and their bonds aren't actually springs, but the model enables people to reason about chemical compounds and their reactions. A more sophisticated model of an atom has a small central ball (the nucleus) surrounded by orbiting electrons. It's also wrong, but the extra complexity enables people to explain more and to solve more problems. (Like software, mental models are never finished: they're just used.)

Presenting a novice with a pile of facts is counter-productive because they don't yet have a model to fit those facts into. In fact, presenting too many facts too soon can actually reinforce the incorrect mental model they've cobbled together. As [Mull2007a] observed in a study of video instruction for science students:

> Students have existing ideas about... phenomena before viewing a video. If the video presents... concepts in a clear, well illustrated way, students believe they are learning but they do not engage with the media on a deep enough level to realize that what is presented differs from their prior knowledge... There is hope, however. Presenting students' common misconceptions in a video alongside the... concepts has been shown to increase learning by increasing the amount of mental effort students expend while watching it.

Your goal when teaching novices should therefore be to help them construct a mental model so that they have somewhere to put facts. For example, Software Carpentry's lesson on the Unix shell[1] introduces fifteen commands in three hours. That's one command every twelve minutes, which seems glacially slow until you realize that the lesson's real purpose isn't to teach those fifteen commands: it's to teach paths, history, tab completion, wildcards, pipes, command-line arguments and redirection. Specific commands don't make sense until novices understand those concepts; once they do, they can start to read manual pages, search for the right keywords on the web and tell whether the results of their searches are useful or not.

The cognitive differences between novices and competent practitioners underpin the differences between two kinds of teaching materials. A **tutorial** helps newcomers to field build a mental model; a **manual**, on the other hand, helps competent practitioners fill in the gaps in their knowledge. Tutorials frustrate competent practitioners because they move too slowly and say things that are obvious (though they are anything *but* obvious to novices). Equally, manuals frustrate novices because they use jargon and *don't* explain things. This phenomenon is called the **expertise reversal effect** [Kaly2003], and is one of the reasons you have to decide early on who your lessons are for.

A Handful of Exceptions

One of the reasons Unix and C became popular is that [Kern1978; Kern1983; Kern1988] somehow managed to be good tutorials and good manuals at the same time. [Fehi2008] and [Ray2014] are among the very few other books in computing that achieve this; even after re-reading them several times, I don't know how they pull it off.

2.1 ARE PEOPLE LEARNING?

Mark Twain once wrote, "It ain't what you don't know that gets you into trouble. It's what you know for sure that just ain't so." One of the exercises in building a mental model is therefore to clear away things that *don't* belong. Broadly speaking, novices' misconceptions fall into three categories:

Factual errors like believing that Vancouver is the capital of British Columbia (it's Victoria). These are usually simple to correct.

Broken models like believing that motion and acceleration must be in the same direction. We can address these by having novices reason through examples where their models give the wrong answer.

Fundamental beliefs such as "the world is only a few thousand years old" or "some kinds of people are just naturally better at programming than others" [Guzd2015b; Pati2016]. These error are often deeply connected to the learner's social identity, so they resist evidence and rationalize contradictions.

---

[1]http://swcarpentry.github.io/shell-novice/

People learn fastest when teachers identify and clear up learners' misconceptions as the lesson is being delivered. This is called **formative assessment** because it forms (or shapes) the teaching while it is taking place. Learners don't pass or fail formative assessment; instead, it gives both the teacher and the learner feedback on how well they are doing and what they should focus on next. For example, a music teacher might ask a learner to play a scale very slowly to check their breathing. The learner finds out if they are breathing correctly, while the teacher gets feedback on whether the explanation they just gave made sense.

Summing Up

*The counterpoint to formative assessment is **summative assessment**, which takes place at the end of the lesson. Summative assessment is like a driver's test: it tells the learner whether they have mastered the topic and the teacher whether their lesson was successful. One way of thinking about the difference is that a chef tasting food as she cooks it is formative assessments, but the guests tasting it once it's served is summative.*

Unfortunately, school has trained most people to believe that all assessment is summative, i.e., that if something feels like a test, doing poorly will count against you. Making formative assessments feel informal helps reduce this anxiety; in my experience, using online quizzes, clickers, or anything else seems to increase it, since most people today believe that anything they do on the web is being watched and recorded.

In order to be useful during teaching, a formative assessment has to be quick to administer (so that it doesn't break the flow of the lesson) and have an unambiguous correct answer (so that it can be used with groups). The most widely used kind of formative assessment is probably the multiple choice question (MCQ). A lot of teachers have a low opinion of them, but when they are designed well, they can reveal much more than just whether someone knows specific facts. For example, suppose you are teaching children how to do multi-digit addition [Ojos2015] and you give them this MCQ:

What is 37 + 15?
a) 52
b) 42
c) 412
d) 43

The correct answer is 52, but the other answers provide valuable insights:

- If the child chooses 42, she has no understanding of what "carrying" means. (She might well write 12 as the answers to 7+5, then overwrite the 1 with the 4 she gets from 3+1.)
- If she chooses 412, she is treating each column of numbers as a separate problem. This is still wrong, but it's wrong for a different reason.

• If she chooses 43 then she knows she has to carry the 1 but is carrying it back into the column it came from. Again, this is a different mistake, and requires a different clarifying explanation from the teacher.

Each of these incorrect answers is a **plausible distractor** with **diagnostic power**. A distractor is a wrong or less-than-best answer; "plausible" means that it looks like it could be right, while "diagnostic power" means that each of the distractors helps the teacher figure out what to explain next to that particular learner.

The spread of responses to a formative assessment guides what you do next. If enough of the class has the right answer, you move on. If the majority of the class chooses the same wrong answer, you should go back and work on correcting the misconception that distractor points to. If their answers are evenly split between several options they are probably just guessing, so you should back up and re-explain the idea in a different way. (Repeating exactly the same explanation will probably not be useful, which is one of things that makes so many video courses pedagogically ineffective.)

What if most of the class votes for the right answer but a few vote for wrong ones? In that case, you have to decide whether you should spend time getting the minority caught up or whether it's more important to keep the majority engaged. No matter how hard you work or what teaching practices you use, you won't always be able to give everyone what they need; it's your responsibility as a teacher to make the call.

Where Do Wrong Answers Come From?

In order to come up with plausible distractors, think about the questions your learners asked or problems they had the last time you taught this subject. If you haven't taught it before, think about your own misconceptions, ask colleagues about their experiences, or look at the history of your field: if everyone misunderstood your subject in some way fifty years ago, the odds are that a lot of your learners will still misunderstand it that way today. You can also ask open-ended questions in class to collect misconceptions about material to be covered in a later class, or check question and answer sites like Quora[2] or Stack Overflow[3] to see what people learning the subject elsewhere are confused by.

Developing formative assessments makes your lessons better because it forces you to think about your learners' mental models. In my experience, once I do this I automatically write the lesson to cover the most likely gaps and errors. Formative assessments therefore pay off even if they aren't used (though teaching is more effective if they are).

MCQs aren't the only kind of formative assessment: Chapter 12 describes other kinds of exercises that are quick and unambiguous. Whatever you pick, you should do something that takes a minute or two every 10–15 minutes to make sure that your

---

[2]http://www.quora.com

[3]https://stackoverflow.com/

learners are actually learning. This rhythm isn't based on an intrinsic attentional limit: [Wils2007] found little support for the often-repeated claim that learners can only pay attention for 10–15 minutes. Instead, the guideline ensures that if a significant number of people have fallen behind, you only have to repeat a short portion of the lesson. Frequent formative assessments also keep learners engaged, particularly if they involved small-group discussion (Section 9.2).

Formative assessments can also be used *before* lessons. If you start a class with an MCQ and everyone answers it correctly, you can avoid explaining something that your learners already know. This kind of **active teaching** gives you more time to focus on things they don't know. It also shows learners that you respect their time enough not to waste it, which helps with motivation (Chapter 10).

Concept Inventories

Given enough data, MCQs can be made surprisingly precise. The best-known example is the **Force Concept Inventory** *[Hest1992], which assesses understanding of basic Newtonian mechanics. By interviewing a large number of respondents, correlating their misconceptions with patterns of right and wrong answers, and then improving the questions, its creators constructed a diagnostic tool that can pinpoint specific misconceptions. Researchers can then use that tool to measure the effect of changes in teaching methods [Hake1998].*

Tew and others developed and validated a language-independent assessment for introductory programming [Tew2011]; [Park2016] replicated it, and [Hamo2017] is developing a concept inventory for recursion. However, it's very costly to build tools like this, and learners' ability to search for answers online is an ever-increasing threat to their validity.

Working formative assessments into class only requires a little bit of preparation and practice. Giving learners colored or numbered cards so that they can all answer an MCQ at once (rather than holding up their hands in turn), having one of the options be, "I have no idea," and encouraging them to talk to their neighbors for a few seconds before answering will all help ensure that your teaching flow isn't disrupted. Section 9.2 describes a powerful, evidence-based teaching method that builds on these simple ideas.

Humor

Teachers sometimes put supposedlysilly answers like "my nose!" on MCQs, particularly ones intended for younger learners. However, these don't provide any insight into learners' misconceptions, and most people don't actually find them funny. As a rule, you should only include a joke in a lesson if you find it funny the third time you re-read it.

A lesson's formative assessments should prepare learners for its summative assessment: no one should ever encounter a question on an exam that the teaching did not prepare them for. This doesn't mean you should never put new kinds of problems on an exam, but if you do, you should have given learners practice tackling novel problems beforehand. Chapter 6 explores this in depth.

2.2 NOTIONAL MACHINES

The term **computational thinking** is bandied about a lot, in part because people can agree it's important while meaning very different things by it. Rather than arguing over what it does and doesn't include, it's more useful to think about the **notional machine** that you want learners to understand [DuBo1986]. According to [Sorv2013], a notional machine:

- is an idealized abstraction of computer hardware and other aspects of programs' runtime environments;
- enables the semantics of programs to be described; and
- correctly reflects what programs do when executed.

For example, my notional machine for Python is:

1. Running programs live in memory, which is divided between a call stack and a heap.
2. Memory for data is always allocated from the heap.
3. Every piece of data is stored in a two-part structure. The first part says what type the data is, and the second part is the actual value.
4. Booleans, numbers and character strings are never modified after they are created.
5. Lists, sets and other collections store references to other data rather than storing those values directly. They can be modified after they are created, i.e., a list can be extended or new values can be added to a set.
6. When code is loaded into memory, Python converts it to a sequence of instructions that are stored like any other data. This is why it's possible to assign functions to variables and pass them as parameters.
7. When code is executed, Python steps through the instructions, doing what each one tells it to in turn.
8. Some instructions make Python read data, do calculations, and create new data. Other instructions control what instructions Python executes, which is how loops and conditionals work. Yet another instruction tells Python to call a function.
9. When a function is called, Python pushes a new stack frame onto the call stack.
10. Each stack frame stores variables' names and references to data. Function parameters are just another kind of variable.
11. When a variable is used, Python looks for it in the top stack frame. If it isn't there, it looks in the bottom (global) frame.
12. When the function finishes, Python erases its stack frame and jumps back to the instructions it was executing before the function call. If there isn't a "before," the program has finished.

I use this cartoon version of reality whenever I teach Python. After about 25 hours of instructions and 100 hours of work on their own time, I expect most learners to have a mental model that includes most or all of these features.

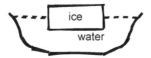

Figure 2.1: Ice in a bathtub

2.3 EXERCISES

YOUR MENTAL MODELS (THINK-PAIR-SHARE/15)

What is one mental model you use to understand your work? Write a few sentences describing it and give feedback on a partner's. Once you have done that, have a few people share their models with the whole group. Does everyone agree on what a mental model is? Is it possible to give a precise definition, or is the concept useful precisely because it is fuzzy?

SYMPTOMS OF BEING A NOVICE (WHOLE CLASS/5)

Saying that novices don't have a mental model of a particular domain is not the same as saying that they don't have a mental model at all. Novices tend to reason by analogy and guesswork, borrowing bits and pieces of mental models from other domains that seem superficially similar.

People who are doing this often say things that are not even wrong[4]. As a class, discuss what some other symptoms of being a novice are. What does someone do or say that leads you to classify them as a novice in some domain?

MODELING NOVICE MENTAL MODELS (PAIRS/20)

Create a multiple choice question related to a topic you have taught or intend to teach and explain the diagnostic power of each its distractors (i.e., what misconception each distractor is meant to identify).

When you are done, trade MCQs with a partner. Is their question ambiguous? Are the misconceptions plausible? Do the distractors actually test for them? Are any likely misconceptions *not* tested for?

THINKING THINGS THROUGH (WHOLE CLASS/15)

A good formative assessment requires people to think through a problem. For example, imagine that you have placed a block of ice in a bathtub and then filled the tub to the rim with water. When the ice melts, does the water level go up (so that the tub overflows), go down, or stay the same (Figure 2.1)?

The correct answer is that the level stays the same: the ice displaces its own weight in water, so it exactly fills the "hole" it has made when it melts. Figuring out why

---

[4]https://en.wikipedia.org/wiki/Not_even_wrong

helps people build a model of the relationship between weight, volume and density [Epst2002].

Describe another formative assessment you have seen or used that required people to think something through and thereby identify flaws in their reasoning. When you are done, explain your example to a partner and give them feedback on theirs.

A DIFFERENT PROGRESSION (INDIVIDUAL/15)

The novice-competent-expert model of skill development is sometimes called the Dreyfus model[5]. Another commonly used progression is the four stages of competence[6]:

Unconscious incompetence: the person doesn't know what they don't know.

Conscious incompetence: the person realizes that they don't know something.

Conscious competence: the person has learned how to do something, but can only do it while concentrating and may still need to break things down into steps.

Unconscious competence: the skill has become second nature and the person can do it reflexively.

Identify one subject where you are at each level. What level are most of your learners at in the subject you teach most often? What level are you trying to get them to? How do these four stages relate to the novice-competent-expert classification?

WHAT KIND OF COMPUTING? (INDIVIDUAL/10)

[Tedr2008] summarizes three traditions in computing:

Mathematical: Programs are the embodiment of algorithms. They are either correct or incorrect, as well as more or less efficient.

Scientific: Programs are more or less accurate models of information processes that can be studied using the scientific method.

Engineering: Programs are built objects like dams and airplanes and are more or less effective and reliable.

Which of these best matches your mental model of computing? If none of them do, what model do you have?

---

[5]https://en.wikipedia.org/wiki/Dreyfus_model_of_skill_acquisition

[6]https://en.wikipedia.org/wiki/Four_stages_of_competence

EXPLAINING WHY NOT (PAIRS/5)

One of your learners thinks that there is some kind of difference between text that they type in character by character and identical text that they copy and paste. Think of a reason they might believe this or something that might have happened to give them this impression, then pretend to be that learner while your partner explains why this isn't the case. Trade roles and try again.

YOUR MODEL NOW (WHOLE CLASS/5)

As a class, create a list of the key elements of your mental model of learning. What are the half-dozen most important concepts and how do they relate?

YOUR NOTIONAL MACHINES (SMALL GROUPS/20)

Working in small groups, write up a description of the notional machine you want learners to use to understand how their programs run. How does a notional machine for a blocks-based language like Scratch differ from that for Python? What about a notional machine for spreadsheets or for a browser that is interpreting HTML and CSS when rendering a web page?

ENJOYING WITHOUT LEARNING (INDIVIDUAL/5)

Multiple studies have shown that teaching evaluations don't correlate with learning outcomes [Star2014; Uttl2017], i.e., that how highly learners rate a course doesn't predict how much they remember. Have you ever enjoyed a class that you didn't actually learn anything from? If so, what made it enjoyable?

REVIEW

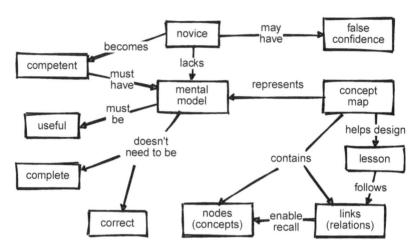

Figure 2.2: Concepts: Mental models

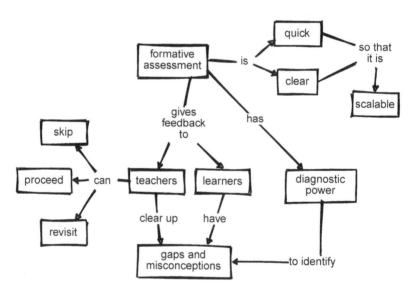

Figure 2.3: Concepts: Assessment

3 Expertise and Memory

Memory is the residue of thought.
— Daniel Willingham

The previous chapter explained the differences between novices and competent practitioners. This one looks at expertise: what it is, how people acquire it, and how it can be harmful as well as helpful. We then introduce one of the most important limits on learning and look at how drawing pictures of mental models can help us turn knowledge into lessons.

To start, what do we mean when we say someone is an expert? The usual answer is that they can solve problems much faster than people who are "merely competent", or that they can recognize and deal with cases where the normal rules don't apply. They also somehow make this look effortless: in many cases, they seem to know the right answer at a glance [Parn2017].

Expertise is more than just knowing more facts: competent practitioners can memorize a lot of trivia without noticeably improving their performance. Instead, imagine for a moment that we store knowledge as a network or graph in which facts are nodes and relationships are arcs[1]. The key difference between experts and competent practitioners is that experts' mental models are much more densely connected, i.e., they are more likely to know a connection between any two facts.

The graph metaphor explains why helping learners make connections is as important as introducing them to facts: without those connections, people can't recall and use what they know. It also explains many observed aspects of expert behavior:

- Experts can often jump directly from a problem to a solution because there actually is a direct link between the two in their mind. Where a competent practitioner would have to reason $A{\rightarrow}B{\rightarrow}C{\rightarrow}D{\rightarrow}E$, an expert can go from A to E in a single step. We call this **intuition**: instead of reasoning their way to a solution, the expert recognizes a solution in the same way that they would recognize a familiar face.
- Densely connected graphs are also the basis for experts' **fluid representations**, i.e., their ability to switch back and forth between different views of a problem [Petr2016]. For example, when trying to solve a problem in mathematics, an expert might switch between tackling it geometrically and representing it as a set of equations.
- This metaphor also explains why experts are better at diagnosis than competent practitioners: more linkages between facts makes it easier to reason backward from symptoms to causes. (This in turn is why asking programmers to debug during job interviews gives a more accurate impression of their ability than asking them to program.)

---

[1] This is definitely *not* how our brains work, but it's a useful metaphor.

• Finally, experts are often so familiar with their subject that they can no longer imagine what it's like to *not* see the world that way. This means they are often less able to teach the subject than people with less expertise who still remember learning it themselves.

The last of these points is called **expert blind spot**. As originally defined in [Nath2003], it is the tendency of experts to organize explanation according to the subject's deep principles rather than being guided by what their learners already know. It can be overcome with training, but it is part of reason there is no correlation between how good someone is at doing research in an area and how good they are at teaching it [Mars2002].

The J Word

Experts often betray their blind spot by using the word "just," as in, "Oh, it's easy, you just fire up a new virtual machine and then you just install these four patches to Ubuntu and then you just re-write your entire program in a pure functional language." As we discuss in Chapter 10, doing this signals that the speaker thinks the problem is trivial and that the person struggling with it must therefore be stupid, so don't do this.

3.1 CONCEPT MAPS

Our tool of choice for representing someone's mental model is a **concept map**, in which facts are bubbles and connections are labeled connections. As examples, Figure 3.1 shows why the Earth has seasons (from IHMC[2]), and Appendix G presents concept maps for libraries from three points of view.

To show how concept maps can be using in teaching programming, consider this for loop in Python:

```
for letter in "abc":
    print(letter)
```

whose output is:

```
a
b
c
```

The three key "things" in this loop are shown in the top of Figure 3.2, but they are only half the story. The expanded version in the bottom shows the relationships between those things, which are as important for understanding as the concepts themselves.

---

[2]https://cmap.ihmc.us/

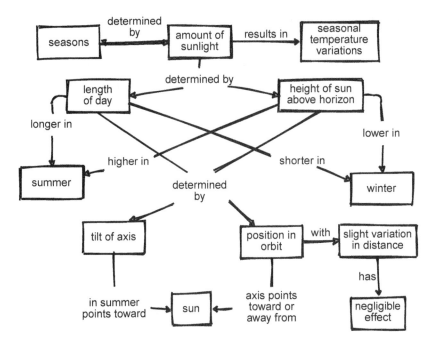

Figure 3.1: Concept map for seasons

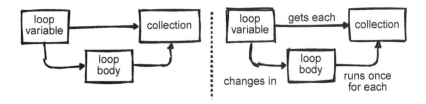

Figure 3.2: Concept map for a for loop

Concept maps can be used in many ways:

Helping teachers figure out what they're trying to teach. A concept map separates content from order: in our experience, people rarely wind up teaching things in the order in which they first drew them.

Aiding communication between lesson designers. Teachers with very different ideas of what they're trying to teach are likely to pull their learners in different directions. Drawing and sharing concept maps can help prevent this. And yes, different people may have different concept maps for the same topic, but concept mapping makes those differences explicit.

Aiding communication with learners. While it's possible to give learners a predrawn map at the start of a lesson for them to annotate, it's better to draw it piece by piece while teaching to reinforce the ties between what's in the map and what the teacher said. We will return to this idea in Section 4.1.

For assessment. Having learners draw pictures of what they think they just learned shows the teacher what they missed and what was miscommunicated. Reviewing learners' concept maps is too time-consuming to do as a formative assessment during class, but very useful in weekly lectures *once learners are familiar with the technique*. The qualification is necessary because any new way of doing things initially slows people down—if a learner is trying to make sense of basic programming, asking them to figure out how to draw their thoughts at the same time is an unfair load.

Some teachers are also skeptical of whether novices can effectively map their understanding, since introspection and explanation of understanding are generally more advanced skills than understanding itself. For example, [Kepp2008] looked at the use of concept mapping in computing education. One of their findings was that, "...concept mapping is troublesome for many students because it tests personal understanding rather than knowledge that was merely learned by rote." As someone who values understanding over rote knowledge, I consider that a benefit.

Start Anywhere
When first asked to draw a concept map, many people will not know where to start. When this happens, write down two words associated with the topic you're trying to map, then draw a line between them and add a label explaining how those two ideas are related. You can then ask what other things are related in the same way, what parts those things have, or what happens before or after the concepts already on the page in order to discover more nodes and arcs. After that, the hard part is often stopping.

Concept maps are just one way to represent our understanding of a subject [Eppl2006]; others include Venn diagrams, flowcharts, and decision trees [Abel2009]. All of these **externalize cognition**, i.e., make mental models visible so that they can be compared and combined[3].

---

[3]To paraphrase Oscar Wilde's Lady Windermere, people often don't know what they're thinking until they've heard themselves say it.

Rough Work and Honesty

Many user interface designers believe that it's better to show people rough sketches of their ideas rather than polished mock-ups because people are more likely to give honest feedback on something that they think only took a few minutes to create: if it looks as though what they're critiquing took hours to create, most will pull their punches. When drawing concept maps to motivate discussion, you should therefore use pencils and scrap paper (or pens and a whiteboard) rather than fancy computer drawing tools.

3.2   SEVEN PLUS OR MINUS TWO

While the graph model of knowledge is wrong but useful, another simple model has a sounder physiological basis. As a rough approximation, human memory can be divided into two distinct layers. The first, called **long-term** or **persistent memory**, is where we store things like our friends' names, our home address, and what the clown did at our eighth birthday party that scared us so much. Its capacity is essentially unlimited, but it is slow to access—too slow to help us cope with hungry lions and disgruntled family members.

Evolution has therefore given us a second system called **short-term** or **working memory**. It is much faster, but also much smaller: [Mill1956] estimated that the average adult's working memory could only hold 7±2 items at a time. This is why phone numbers[4] are 7 or 8 digits long: back when phones had dials instead of keypads, that was the longest string of numbers most adults could remember accurately for as long as it took the dial to go around several times.

Participation

The size of working memory is sometimes used to explain why sports teams tend to have about half a dozen members or are broken into sub-groups like the forwards and backs in rugby. It is also used to explain why meetings are only productive up to a certain number of participants: if twenty people try to discuss something, either three meetings are going on at once or half a dozen people are talking while everyone else listens. The argument is that people's ability to keep track of their peers is constrained by the size of working memory, but so far as I know, the link has never been proven.

7±2 is the single most important number in teaching. A teacher cannot place information directly in a learner's long-term memory. Instead, whatever they present is first stored in the learner's short-term memory, and is only transferred to long-term memory after it has been held there and rehearsed (Section 5.1). If the teacher presents too much information too quickly, the new information displaces the old before the latter is transferred.

---

[4]https://www.quora.com/Why-did-Bell-Labs-create-phone-numbers-of-7-digits-10-digits-Is-there-a-reason-that-dashes-and-brackets-are-used

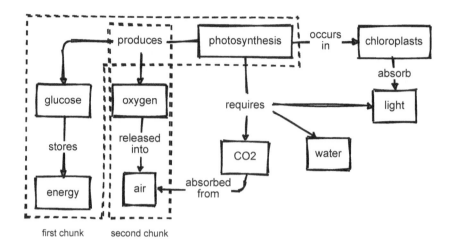

Figure 3.3: Using concept maps in lesson design

This is one of the ways to use a concept map when designing a lesson: it helps make sure learners' short-term memories won't be overloaded. Once the map is drawn, the teacher chooses a subsection that will fit in short-term memory and lead to a formative assessment (Figure 3.3), then adds another subsection for the next lesson episode and so on.

Building Concept Maps Together

The next time you have a team meeting, give everyone a sheet of paper and have them spend a few minutes drawing their own concept map of the project you're all working on. On the count of three, have everyone reveal their concept maps to their group. The discussion that follows may help people understand why they've been tripping over each other.

Note that the simple model of memory presented here has largely been replaced by a more sophisticated one in which short-term memory is broken down into several modal stores (e.g., for visual vs. linguistic memory), each of which does some involuntary preprocessing [Mill2016a]. Our presentation is therefore an example of a mental model that aids learning and everyday work.

PATTERN RECOGNITION

Recent research suggests that the actual size of short-term memory might be as low as 4 ± 1 items [Dida2016]. In order to handle larger sets of information, our minds create **chunks**. For example, most of us remember words as single items rather than as sequences of letters. Similarly, the pattern made by five spots on cards or dice is remembered as a whole rather than as five separate pieces of information.

Experts have more and larger chunks than non-experts, i.e., experts "see" larger patterns and have more patterns to match things against. This allows them to reason at a higher level and to search for information more quickly and more accurately. However, chunking can also mislead us if we mis-identify things: newcomers really can sometimes see things that experts have looked at and missed.

Given how important chunking is to thinking, it is tempting to identify design patterns[5] and teach them directly. These patterns help competent practitioners think and talk to each other in many domains (including teaching [Berg2012]), but pattern catalogs are too dry and too abstract for novices to make sense of on their own. That said, giving names to a small number of patterns does seem to help with teaching, primarily by giving the learners a richer vocabulary to think and communicate with [Kuit2004; Byck2005; Saja2006]. We will return to this in Section 7.5.

3.3 BECOMING AN EXPERT

So how does someone become an expert? The idea that ten thousand hours of practice will do it is widely quoted but probably not true[6]: doing the same thing over and over again is much more likely to solidify bad habits than improve performance. What actually works is doing similar but subtly different things, paying attention to what works and what doesn't, and then changing behavior in response to that feedback to get cumulatively better. This is called **deliberate** or **reflective practice**, and a common progression is for people to go through three stages:

Act on feedback from others. A learner might write an essay about what they did on their summer holiday and get feedback from a teacher telling them how to improve it.

Give feedback on others' work. The learner might critique character development in a Harry Potter novel and get feedback from the teacher on their critique.

Give feedback to themselves. At some point, the learner starts critiquing their own work as they do it using the skills they have now built up. Doing this is so much faster than waiting for feedback from others that proficiency suddenly starts to take off.

> ### What Counts as Deliberate Practice?
> *[Macn2014] found that, "... deliberate practice explained 26% of the variance in performance for games, 21% for music, 18% for sports, 4% for education, and less than 1% for professions." However, [Eric2016] critiqued this finding by saying, "Summing up every hour of any type of practice during an individual's career implies that the impact of all types of practice activity on performance is equal—an assumption that... is inconsistent with the evidence." To be effective, deliberate practice requires both a clear performance goal and immediate informative feedback, both of which are things teachers should strive for anyway.*

---

[5] https://en.wikipedia.org/wiki/Software_design_pattern

[6] http://www.goodlifeproject.com/podcast/anders-ericsson/

3.4 EXERCISES

CONCEPT MAPPING (PAIRS/30)

Draw a concept map for something you would teach in five minutes. Trade with a
partner and critique each other's maps. Do they present concepts or surface detail?
Which of the relationships in your partner's map do you consider concepts and vice
versa?

CONCEPT MAPPING (AGAIN) (SMALL GROUPS/20)

Working in groups of 3–4, have each person independently draw a concept map
showing their mental model of what goes on in a classroom. When everyone is done,
compare the concept maps. Where do your mental models agree and disagree?

ENHANCING SHORT-TERM MEMORY (INDIVIDUAL/5 MINUTES)

[Cher2007] suggests that the main reason people draw diagrams when they are discussing things is to enlarge their short-term memory: pointing at a wiggly bubble
drawn a few minutes ago triggers recall of several minutes of debate. When you
exchanged concept maps in the previous exercise, how easy was it for other people
to understand what your map meant? How easy would it be for you if you set it aside
for a day or two and then looked at it again?

THAT'S A BIT SELF-REFERENTIAL, ISN'T IT? (WHOLE CLASS/30)

Working independently, draw a concept map for concept maps. Compare your concept map with those drawn by other people. What did most people include? What
were the significant differences?

NOTICING YOUR BLIND SPOT (SMALL GROUPS/10)

Elizabeth Wickes listed all the things you need to understand[7] in order to read this
one line of Python:

```
answers = ['tuatara', 'tuataras', 'bus', "lick"]
```

- The square brackets surrounding the content mean we're working with a list (as
  opposed to square brackets immediately to the right of something, which is a data
  extraction notation).
- The elements are separated by commas outside and between the quotes (rather
  than inside, as they would be for quoted speech).
- Each element is a character string, and we know that because of the quotes. We
  could have number or other data types in here if we wanted; we need quotes
  because we're working with strings.

---

[7]https://twitter.com/elliewix/status/981285432922202113

- We're mixing our use of single and double quotes; Python doesn't care so long as they balance around the individual strings.
- Each comma is followed by a space, which is not required by Python, but which we prefer it for readability.

Each of these details might be overlooked by an expert. Working in groups of 3–4, select something equally short from a lesson you have recently taught or learned and break it down to this level of detail.

WHAT TO TEACH NEXT (INDIVIDUAL/5)

Refer back to the concept map for photosynthesis in Figure 3.3. How many concepts and links are in the selected chunks? What would you include in the next chunk of the lesson and why?

THE POWER OF CHUNKING (INDIVIDUAL/5)

Look at Figure 3.4 for 10 seconds, then look away and try to write out your phone number with these symbols[8]. (Use a space for '0'.) When you are finished, look at the alternative representation in Appendix H. How much easier are the symbols to remember when the pattern is made explicit?

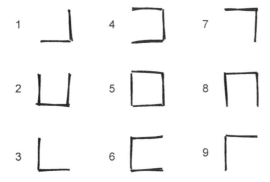

Figure 3.4: Unchunked representation

---

[8]My thanks to Warren Code for introducing me to this example.

4 Cognitive Architecture

We have been talking about mental models as if they were real things, but what actually goes on in a learner's brain when they're learning? The short answer is that we don't know; the longer answer is that we know a lot more than we used to. This chapter will dig a little deeper into what brains do while they're learning and how we can leverage that to design and deliver lessons more effectively.

4.1 WHAT'S GOING ON IN THERE?

Figure 4.1 is a simplified model of human cognitive architecture. The core of this model is the separation between short-term and long-term memory discussed in Section 3.2. Long-term memory is like your basement: it stores things more or less permanently, but you can't access its contents directly. Instead, you rely on your short-term memory, which is the cluttered kitchen table of your mind.

When you need something, your brain retrieves it from long-term memory and puts it in short-term memory. Conversely, new information that arrives in short-term memory has to be encoded to be stored in long-term memory. If that information isn't encoded and stored, it's not remembered and learning hasn't taken place.

Information gets into short-term memory primarily through your verbal channel (for speech) and visual channel (for images)[1]. Most people rely primarily on their visual channel, but when images and words complement each other, the brain does a better job of remembering them both: they are encoded together, so recall of one later on helps trigger recall of the other.

Linguistic and visual input are processed by different parts of the human brain, and linguistic and visual memories are stored separately as well. This means that

---

[1] A more complete model would also include your senses of touch, smell, and taste, but we'll ignore those for now.

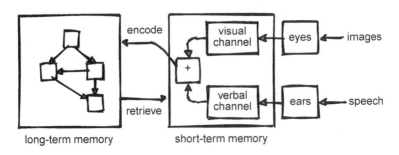

Figure 4.1: Cognitive architecture

correlating linguistic and visual streams of information takes cognitive effort: when someone reads something while hearing it spoken aloud, their brain can't help but check that it's getting the same information on both channels.

Learning is therefore increased when information is presented simultaneously in two different channels, but is reduced when that information is redundant rather than complementary, a phenomenon called the **split-attention effect** [Maye2003]. For example, people generally find it harder to learn from a video that has both narration and on-screen captions than from one that has either the narration or the captions but not both, because some of their attention has to be devoted to checking that the narration and the captions agree with each other. Two notable exceptions to this are people who do not yet speak the language well and people with hearing impairments or other special needs, both of whom may find that the value of the redundant information outweighs the extra processing effort.

Piece by Piece

The split attention effect explains why it's more effective to draw a diagram piece by piece while teaching than to present the whole thing at once. If parts of the diagram appear at the same time as things are being said, the two will be correlated in the learner's memory. Pointing at part of the diagram later is then more likely to trigger recall of what was being said when that part was being drawn.

The split-attention effect does *not* mean that learners shouldn't try to reconcile multiple incoming streams of information—after all, this is what they have to do in the real world [Atki2000]. Instead, it means that instruction shouldn't require people to do it while they are first mastering unit skills; instead, using multiple sources of information simultaneously should be treated as a separate learning task.

Not All Graphics are Created Equal

[Sung2012] presents an elegant study that distinguishes seductive graphics (which are highly interesting but not directly relevant to the instructional goal), decorative graphics (which are neutral but not directly relevant to the instructional goal), and instructive graphics (which are directly relevant to the instructional goal). Learners who received any kind of graphic gave material higher satisfaction ratings than those who didn't get graphics, but only learners who got instructive graphics actually performed better.

Similarly, [Stam2013; Stam2014] found that having more information can actually lower performance. They showed children pictures, pictures and numbers, or just numbers for two tasks. For some, having pictures or pictures and numbers outperformed having numbers only, but for others, having pictures outperformed pictures and numbers, which outperformed just having numbers.

4.2 COGNITIVE LOAD

In [Kirs2006], Kirschner, Sweller and Clark wrote:

> Although unguided or minimally guided instructional approaches are very pop-
> ular and intuitively appealing...these approaches ignore both the structures
> that constitute human cognitive architecture and evidence from empirical stud-
> ies over the past half-century that consistently indicate that minimally guided
> instruction is less effective and less efficient than instructional approaches that
> place a strong emphasis on guidance of the student learning process. The
> advantage of guidance begins to recede only when learners have sufficiently
> high prior knowledge to provide "internal" guidance.

Beneath the jargon, the authors were claiming that having learners ask their own
questions, set their own goals, and find their own path through a subject is less effective than showing them how to do things step by step. The "choose your own adventure" approach is known as **inquiry-based learning**, and is intuitively appealing:
after all, who would argue *against* having learners use their own initiative to solve
real-world problems in realistic ways? However, asking learners to do this in a new
domain overloads them by requiring them to master a domain's factual content and
its problem-solving strategies at the same time.

More specifically, **cognitive load theory** proposed that people have to deal with
three things when they're learning:

Intrinsic load is what people have to keep in mind in order to absorb new material.

Germane Load is the (desirable) mental effort required to link new information to
old, which is one of the things that distinguishes learning from memorization.

Extraneous Load is anything that distracts from learning.

Cognitive load theory holds that people have to divide a fixed amount of working
memory between these three things. Our goal as teachers is to maximize the memory
available to handle intrinsic load, which means reducing the germane load at each
step and eliminating the extraneous load.

PARSONS PROBLEMS

One kind of exercise that can be explained in terms of cognitive load is often used
when teaching languages. Suppose you ask someone to translate the sentence, "How
is her knee today?" into Frisian. To solve the problem, they need to recall both vocabulary and grammar, which is a double cognitive load. If you ask them to put "hoe,"
"har," "is," "hjoed," and "knie" in the right order, on the other hand, you are allowing
them to focus solely on learning grammar. If you write these words in five different
fonts or colors, though, you have increased the extraneous cognitive load, because
they will involuntarily (and possibly unconsciously) expend some effort trying to
figure out if the differences are meaningful (Figure 4.2).

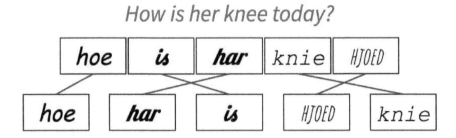

Figure 4.2: Constructing a sentence

The coding equivalent of this is called a **Parsons Problem**[2] [Pars2006]. When teaching people to program, you can give them the lines of code they need to solve a problem and ask them to put them in the right order. This allows them to concentrate on control flow and data dependencies without being distracted by variable naming or trying to remember what functions to call. Multiple studies have shown that Parsons Problems take less time for learners to do but produce equivalent educational outcomes [Eric2017].

FADED EXAMPLES

Another type of exercise that can be explained in terms of cognitive load is to give learners a series of **faded examples**. The first example in a series presents a complete use of a particular problem-solving strategy. The next problem is of the same type, but has some gaps for the learner to fill in. Each successive problem gives the learner less **scaffolding**, until they are asked to solve a complete problem from scratch. When teaching high school algebra, for example, we might start with this:

$$
\begin{aligned}
(4x + 8)/2 &= 5 \\
4x + 8 &= 2 * 5 \\
4x + 8 &= 10 \\
4x &= 10 - 8 \\
4x &= 2 \\
x &= 2 / 4 \\
x &= 1 / 2
\end{aligned}
$$

and then ask learners to solve this:

$$
\begin{aligned}
(3x - 1)*3 &= 12 \\
3x - 1 &= \_ / \_ \\
3x - 1 &= 4 \\
3x &= \_ \\
x &= \_ / 3 \\
x &= \_
\end{aligned}
$$

---

[2]Named after one of its creators.

and this:

$$(5x + 1)*3 \quad = \quad 4$$
$$5x + 1 \quad = \quad \_$$
$$5x \quad = \quad \_$$
$$x \quad = \quad \_$$

and finally this:

$$(2x + 8)/4 \quad = \quad 1$$
$$x \quad = \quad \_$$

A similar exercise for teaching Python might start by showing learners how to find the total length of a list of words:

```
# total_length(["red", "green", "blue"]) => 12
define total_length(list_of_words):
    total = 0
    for word in list_of_words:
        total = total + length(word)
    return total
```

and then ask them to fill in the blanks in this (which focuses their attention on control structures):

```
# word_lengths(["red", "green", "blue"]) => [3, 5, 4]
define word_lengths(list_of_words):
    list_of_lengths = []
    for ____ in ____:
        append(list_of_lengths, ____)
    return list_of_lengths
```

The next problem might be this (which focuses their attention on updating the final result):

```
# join_all(["red", "green", "blue"]) => "redgreenblue"
define join_all(list_of_words):
    joined_words = ____
    for ____ in ____:

        ____

    return joined_words
```

Learners would finally be asked to write an entire function on their own:

```
# make_acronym(["red", "green", "blue"]) => "RGB"
define make_acronym(list_of_words):

    ____
```

Faded examples work because they introduce the problem-solving strategy piece by piece: at each step, learners have one new problem to tackle, which is less intimidating than a blank screen or a blank sheet of paper (Section 9.11). It also encourages learners to think about the similarities and differences between various approaches, which helps create the linkages in their mental models that help retrieval.

The key to constructing a good faded example is to think about the problem-solving strategy it is meant to teach. For example, the programming problems above all use the accumulator design pattern, in which the results of processing items from a collection are repeatedly added to a single variable in some way to create the final result.

Cognitive Apprenticeship

*An alternative model of learning and instruction that also uses scaffolding and fading is **cognitive apprenticeship**, which emphasizes the way in which a master passes on skills and insights to an apprentice. The master provides models of performance and outcomes, then coaches novices by explaining what they are doing and why [Coll1991; Casp2007]. The apprentice reflects on their own problem solving, e.g., by thinking aloud or critiquing their own work, and eventually explores problems of their own choosing.*

This model tells us that teachers should present several examples when presenting a new idea so that learners can see what to generalize, and that we should vary the form of the problem to make it clear what are and aren't superficial features[3]. Problems should be presented in real-world contexts, and we should encourage self-explanation to help learners organize and make sense of what they have just been taught (Section 5.1).

LABELED SUBGOALS

Labeling subgoals means giving names to the steps in a step-by-step description of a problem-solving process. [Marg2016; Morr2016] found that learners with labeled subgoals solved Parsons Problems better than learners without, and the same benefit is seen in other domains [Marg2012]. Returning to the Python example used earlier, the subgoals in finding the total length of a list of words or constructing an acronym are:

1. Create an empty value of the type to be returned.
2. Get the value to be added to the result from the loop variable.
3. Update the result with that value.

Labeling subgoals works because grouping related steps into named chunks (Section 3.2) helps learners distinguish what's generic from what is specific to the problem at hand. It also helps them build a mental model of that kind of problem so that

---

[3]For a long time, I believed that the variable holding the value a function was going to return *had* to be called `result` because my teacher always used that name in examples.

they can solve other problems of that kind, and gives them a natural opportunity for self-explanation (Section 5.1).

MINIMAL MANUALS

The purest application of cognitive load theory may be John Carroll's **minimal manual** [Carr1987; Carr2014]. Its starting point is a quote from a user: "I want to do something, not learn how to do everything." Carroll and colleagues redesigned training to present every idea as a single-page self-contained task: a title describing what the page was about, step-by-step instructions of how to do just one thing (e.g., how to delete a blank line in a text editor), and then several notes on how to recognize and debug common problems. They found that rewriting training materials this way made them shorter overall, and that people using them learned faster. Later studies confirmed that this approach outperformed the traditional approach regardless of prior experience with computers [Lazo1993]. [Carr2014] summarized this work by saying:

> Our "minimalist" designs sought to leverage user initiative and prior knowledge, instead of controlling it through warnings and ordered steps. It emphasized that users typically bring much expertise and insight to this learning, for example, knowledge about the task domain, and that such knowledge could be a resource to instructional designers. Minimalism leveraged episodes of error recognition, diagnosis, and recovery, instead of attempting to merely forestall error. It framed troubleshooting and recovery as learning opportunities instead of as aberrations.

4.3 OTHER MODELS OF LEARNING

Critics of cognitive load theory have sometimes argued that any result can be justified after the fact by labeling things that hurt performance as extraneous load and things that don't as intrinsic or germane. However, instruction based on cognitive load theory is undeniably effective. For example, [Maso2016] redesigned a database course to remove split attention and redundancy effects and to provide worked examples and sub-goals. The new course reduced the exam failure rate by 34% and increased learner satisfaction.

A decade after the publication of [Kirs2006], a growing number of people believe that cognitive load theory and inquiry-based approaches are compatible if viewed in the right way. [Kaly2015] argues that cognitive load theory is basically micromanagement of learning within a broader context that considers things like motivation, while [Kirs2018] extends cognitive load theory to include collaborative aspects of learning. As with [Mark2018] (discussed in Section 5.1), researchers' perspectives may differ, but the practical implementation of their theories often wind up being the same.

One of the challenges in educational research is that what we mean by "learning" turns out to be complicated once you look beyond the standardized Western classroom. Two specific perspectives from **educational psychology** have influenced this book. The one we have used so far is **cognitivism**, which focuses on things like pattern recognition, memory formation, and recall. It is good at answering low-level questions, but generally ignores larger issues like, "What do we mean by 'learning'?" and, "Who gets to decide?" The other is **situated learning**, which focuses on bringing people into a community and recognizes that teaching and learning are always rooted in who we are and who we aspire to be. We will discuss it in more detail in Chapter 13.

The Learning Theories website[4] and [Wibu2016] have good summaries of these and other perspectives. Besides cognitivism, those encountered most frequently include **behaviorism** (which treats education as stimulus/response conditioning), **constructivism** (which considers learning an active process during which learners construct knowledge for themselves), and **connectivism** (which holds that knowledge is distributed, that learning is the process of navigating, growing, and pruning connections, and which emphasizes the social aspects of learning made possible by the Internet). These perspectives can help us organize our thoughts, but in practice, we always have to try new methods in the class, with actual learners, in order to find out how well they balance the many forces in play.

4.4 EXERCISES

CREATE A FADED EXAMPLE (PAIRS/30)

It's very common for programs to count how many things fall into different categories: for example, how many times different colors appear in an image, or how many times different words appear in a paragraph of text.

1. Create a short example (no more than 10 lines of code) that shows people how to do this, and then create a second example that solves a similar problem in a similar way but has a couple of blanks for learners to fill in. How did you decide what to fade out? What would the next example in the series be?
2. Define the audience for your examples. For example, are these beginners who only know some basics programming concepts? Or are these learners with some experience in programming?
3. Show your example to a partner, but do *not* tell them what level you think it is for. Once they have filled in the blanks, ask them to guess the intended level.

If there are people among the trainees who don't program at all, try to place them in different groups and have them play the part of learners for those groups. Alternatively, choose a different problem domain and develop a faded example for it.

---

[4]http://www.learning-theories.com/

CLASSIFYING LOAD (SMALL GROUPS/15)

1. Choose a short lesson that a member of your group has taught or taken recently.
2. Make a point-form list of the ideas, instructions, and explanations it contains.
3. Classify each as intrinsic, germane, or extraneous. What did you all agree on? Where did you disagree and why?

(The exercise "Noticing Your Blind Spot" in Section 3.4 will give you an idea of how detailed your point-form list should be.)

CREATE A PARSONS PROBLEM (PAIRS/20)

Write five or six lines of code that does something useful, jumble them, and ask your partner to put them in order. If you are using an indentation-based language like Python, do not indent any of the lines; if you are using a curly-brace language like Java, do not include any of the curly braces. (If your group includes people who aren't programmers, use a different problem domain, such as making banana bread.)

MINIMAL MANUALS (INDIVIDUAL/20)

Write a one-page guide to doing something that your learners might encounter in one of your classes, such as centering text horizontally or printing a number with a certain number of digits after the decimal point. Try to list at least three or four incorrect behaviors or outcomes the learner might see and include a one- or two-line explanation of why each happens and how to correct it.

COGNITIVE APPRENTICESHIP (PAIRS/15)

Pick a coding problem that you can do in two or three minutes and think aloud as you work through it while your partner asks questions about what you're doing and why. Do not just explain what you're doing, but also why you're doing it, how you know it's the right thing to do, and what alternatives you've considered but discarded. When you are done, swap roles with your partner and repeat the exercise.

WORKED EXAMPLES (PAIRS/15)

Seeing worked examples helps people learn to program faster than just writing lots of code [Skud2014], and deconstructing code by tracing it or debugging it also increases learning [Grif2016]. Working in pairs, go through a 10–15 line piece of code and explain what every statement does and why it is necessary. How long does it take? How many things do you feel you need to explain per line of code?

CRITIQUING GRAPHICS (INDIVIDUAL/30)

[Maye2009; Mill2016a] presents six principles for good teaching graphics:

Signalling: visually highlight the most important points so that they stand out from less-critical material.

Spatial contiguity: place captions as close to the graphics as practical to offset the cost of shifting between the two.

Temporal contiguity: present spoken narration and graphics as close in time as practical. (Presenting both at once is better than presenting them one after another.)

Segmenting: when presenting a long sequence of material or when learners are inexperienced with the subject, break the presentation into short segments and let learners control how quickly they advance from to the next.

Pre-training: if learners don't know the major concepts and terminology used in your presentation, teach just those concepts and terms beforehand.

Modality: people learn better from pictures plus narration than from pictures plus text, unless they are non-native speakers or there are technical words or symbols.

Choose a video of a lesson or talk online that uses slides or other static presentations and rate its graphics as "poor," "average," or "good" according to these six criteria.

REVIEW

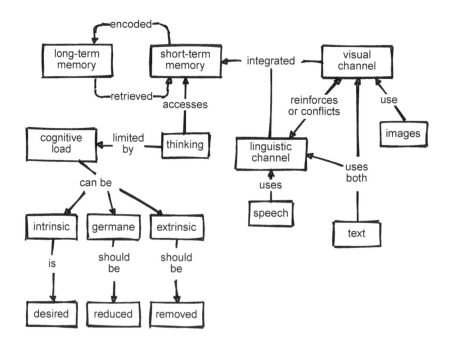

Figure 4.3: Concepts: Cognitive load

5 Individual Learning

Previous chapters have explored what teachers can do to help learners. This chapter looks at what learners can do for themselves by changing their study strategies and getting enough rest.

The most effective strategy is to switch from **passive learning** to **active learning** [Hpl2018], which significantly improves performance and reduces failure rates [Free2014]:

Passive	Active
Read about something	Do exercises
Watch a video	Discuss a topic
Attend a lecture	Try to explain it

Referring back to our simplified model of cognitive architecture (Figure 4.1), active learning is more effective because it keeps new information in short-term memory longer, which increases the odds that it will be encoded successfully and stored in long-term memory. And by using new information as it arrives, learners build or strengthen ties between that information and what they already know, which in turn increases the chances that they will be able to retrieve it later.

The other key to getting more out of learning is **metacognition**, or thinking about one's own thinking. Just as good musicians listen to their own playing and good teachers reflect on their teaching (Chapter 8), learners will learn better and faster if they make plans, set goals, and monitor their progress. It's difficult for learners to master these skills in the abstract—just telling them to make plans doesn't have any effect—but lessons can be designed to encourage good study practices, and drawing attention to these practices in class helps learners realize that learning is a skill they can improve like any other [McGu2015; Miya2018].

The big prize is **transfer of learning**, which occurs when one thing we have learned helps us learn other things more quickly. Researchers distinguish between **near transfer**, which occurs between similar or related areas like fractions and decimals in mathematics, and **far transfer**, which occurs between dissimilar domains— for example, the idea that learning to play chess will help mathematical reasoning or vice versa.

Near transfer undoubtedly occurs—no kind of learning beyond simple memorization could occur if it didn't—and teachers leverage it all the time by giving learners exercises that are similar to material that has just been presented in a lesson. However, [Sala2017] analyzed many studies of far transfer and concluded that:

> …the results show small to moderate effects. However, the effect sizes are inversely related to the quality of the experimental design… We conclude that far transfer of learning rarely occurs.

When far transfer does occur, it seems to happen only once a subject has been mastered [Gick1987]. In practice, this means that learning to program won't help you play chess and vice versa.

5.1 SIX STRATEGIES

Psychologists study learning in a wide variety of ways, but have reached similar conclusions about what actually works [Mark2018]. The Learning Scientists[1] have catalogued six of these strategies and summarized them in a set of downloadable posters[2]. Teaching these strategies to learners, and mentioning them by name when you use them in class, can help them learn how to learn faster and better [Wein2018a; Wein2018b].

SPACED PRACTICE

Ten hours of study spread out over five days is more effective than two five-hour days, and far better than one ten-hour day. You should therefore create a study schedule that spreads study activities over time: block off at least half an hour to study each topic each day rather than trying to cram everything in the night before an exam [Kang2016].

You should also review material after each class, but not immediately after—take at least a half-hour break. When reviewing, be sure to include at least a little bit of older material: for example, spend twenty minutes looking over notes from that day's class and then five minutes each looking over material from the previous day and from a week before. Doing this also helps you catch any gaps or mistakes in previous sets of notes while there's still time to correct them or ask questions: it's painful to realize the night before the exam that you have no idea why you underlined "Demodulate!!" three times.

When reviewing, make notes about things that you had forgotten: for example, make a flash card for each fact that you couldn't remember or that you remembered incorrectly [Matt2019]. This will help you focus the next round of study on things that most need attention.

The Value of Lectures
According to [Mill2016a], "The lectures that predominate in face-to-face courses are relatively ineffective ways to teach, but they probably contribute to spacing material over time, because they unfold in a set schedule over time. In contrast, depending on how the courses are set up, online students can sometimes avoid exposure to material altogether until an assignment is nigh."

---

[1] http://www.learningscientists.org/

[2] http://www.learningscientists.org/downloadable-materials

RETRIEVAL PRACTICE

The limiting factor for long-term memory is not retention (what is stored) but recall (what can be accessed). Recall of specific information improves with practice, so outcomes in real situations can be improved by taking practice tests or summarizing the details of a topic from memory and then checking what was and wasn't remembered. For example, [Karp2008] found that repeated testing improved recall of word lists from 35% to 80%.

Recall is better when practice uses activities similar to those used in testing. For example, writing personal journal entries helps with multiple-choice quizzes, but less than doing practice quizzes [Mill2016a]. This phenomenon is called **transfer-appropriate processing**.

One way to exercise retrieval skills is to solve problems twice. The first time, do it entirely from memory without notes or discussion with peers. After grading your own work against a rubric supplied by the teacher, solve the problem again using whatever resources you want. The difference between the two shows you how well you were able to retrieve and apply knowledge.

Another method (mentioned above) is to create flash cards. Physical cards have a question or other prompt on one side and the answer on the other, and many flash card apps are available for phones. If you are studying as part of a group, swapping flash cards with a partner helps you discover important ideas that you may have missed or misunderstood.

Read-cover-retrieve is a quick alternative to flash cards. As you read something, cover up key terms or sections with small sticky notes. When you are done, go through it a second time and see how well you can guess what's under each of those stickies. Whatever method you use, don't just practice recalling facts and definitions: make sure you also check your understanding of big ideas and the connections between them. Sketching a concept map and then comparing it to your notes or to a previously-drawn concept map is a quick way to do this.

Hypercorrection

One powerful finding in learning research is the **hypercorrection effect** *[Metc2016]. Most people don't like to be told they're wrong, so it would be reasonable to assume that the more confident someone is in the answer they've given on a test, the harder it is to change their mind if they were actually wrong. It turns out that the opposite is true: the more confident someone is that they were right, the more likely they are not to repeat the error if they are corrected.*

INTERLEAVING

One way you can space your practice is to interleave study of different topics: instead of mastering one subject, then a second and third, shuffle study sessions. Even better, switch up the order: A-B-C-B-A-C is better than A-B-C-A-B-C, which in turn is better than A-A-B-B-C-C [Rohr2015]. This works because interleaving fosters creation of more links between different topics, which in turn improves recall.

How long you should spend on each item depends on the subject and how well you know it. Somewhere between 10 and 30 minutes is long enough for you to get into a state of flow (Section 5.2) but not for your mind to wander. Interleaving study will initially feel harder than focusing on one topic at a time, but that's a sign that it's working. If you are using flash cards or practice tests to gauge your progress, you should see improvement after only a couple of days.

ELABORATION

Explaining things to yourself as you go through them helps you understand and remember them. One way to do this is to follow up each answer on a practice quiz with an explanation of why that answer is correct, or conversely with an explanation of why some other plausible answer isn't. Another is to tell yourself how a new idea is similar to or different from one you have seen previously.

Talking to yourself may seem like an odd way to study, but [Biel1995] found that people trained in self-explanation outperformed those who hadn't been trained. Similarly, [Chi1989] found that some learners simply halt when they hit a step they don't understand when trying to solve problems. Others pause and generate an explanation of what's going on, and the latter group learns faster. An exercise to build this skill is to go through an example program line by line with a class, having a different person explain each line in turn and say why it is there and what it accomplishes.

CONCRETE EXAMPLES

One particularly useful form of elaboration is the use of concrete examples. Whenever you have a statement of a general principle, try to provide one or more examples of its use, or conversely take each particular problem and list the general principles it embodies. [Raws2014] found that interleaving examples and definitions like this made it more likely that learners would remember the latter correctly.

One structured way to do this is the ADEPT method[3]: give an **A**nalogy, draw a **D**iagram, present an **E**xample, describe the idea in **P**lain language, and then give the **T**echnical details. Again, if you are studying with a partner or in a group, you can swap and check work: see if you agree that other people's examples actually embody the principle being discussed or which principles are used in an example that they haven't listed.

Another useful technique is to teach by contrast, i.e., to show learners what a solution is *not* or what kind of problem a technique *won't* solve. For example, when showing children how to simplify fractions, it's important to give them a few like 5/7 that can't be simplified so that they don't become frustrated looking for answers that don't exist.

---

[3]https://betterexplained.com/articles/adept-method/

DUAL CODING

The last of the six core strategies that the Learning Scientists[4] describe is to present words and images together. As discussed in Section 4.1, different subsystems in our brains handle and store linguistic and visual information, so if complementary information is presented through both channels, they can reinforce one another. However, learning is less effective when the same information is presented simultaneously in two different channels, because then the brain has to expend effort to check the channels against each other [Maye2003].

One way to take advantage of dual coding is to draw or label timelines, maps, family trees, or whatever else seems appropriate to the material. (I am personally fond of pictures showing which functions call which others in a program.) Drawing a diagram *without* labels, then coming back later to label it, is excellent retrieval practice.

5.2 TIME MANAGEMENT

I used to brag about the hours I was working. Not in so many words, of course—I had *some* social skills—but I would show up for class around noon, unshaven and yawning, and casually mention to whoever would listen that I'd been up working until 6:00 a.m.

Looking back, I can't remember who I was trying to impress. What I remember instead is how much of the work I did in those all-nighters I threw away once I'd had some sleep, and how much damage the stuff I didn't throw away did to my grades.

My mistake was to confuse "working" with "being productive." You can't produce software (or anything else) without doing some work, but you can easily do lots of work without producing anything of value. Convincing people of this can be hard, especially when they're in their teens or twenties, but it pays tremendous dividends.

Scientific study of overwork and sleep deprivation goes back to at least the 1890s—see [Robi2005] for a short, readable summary. The most important results for learners are:

1. Working more than 8 hours a day for an extended period of time lowers your total productivity, not just your hourly productivity—i.e., you get less done in total (not just per hour) when you're in crunch mode.
2. Working over 21 hours in a stretch increases the odds of you making a catastrophic error just as much as being legally drunk.
3. Productivity varies over the course of the workday, with the greatest productivity occurring in the first 4 to 6 hours. After enough hours, productivity approaches zero; eventually it becomes negative.

These facts have been reproduced and verified for over a century, and the data behind them is as solid as the data linking smoking to lung cancer. The problem is

---

[4]http://www.learningscientists.org/

that *people usually don't notice their abilities declining.* Like drunks who think they are still able to drive, people who are deprived of sleep don't realize that they are not finishing their sentences (or thoughts). Five 8-hour days per week has been proven to maximize long-term total output in every industry that has ever been studied; studying or programming are no different.

But what about short bursts now and then, like pulling an all-nighter to meet a deadline? That has been studied too, and the results aren't pleasant. Your ability to think drops by 25% for each 24 hours you're awake. Put it another way, the average person's IQ is only 75 after one all-nighter, which puts them in the bottom 5% of the population. If you do two all-nighters in a row your effective IQ is 50, which is the level at which people are usually judged incapable of independent living.

When You Just Can't Say No

*Research has shown that our ability to exert willpower runs out just like our ability to use muscles: if we have to resist eating the last donut on the tray when we're hungry, we are less likely to fold laundry and vice versa. This is called **ego depletion** [Mill2016a], and an effective counter is to build up habits so that doing the right thing is automatic.*

"But—but—I have so many assignments to do!" you say. "And they're all due at once! I *have* to work extra hours to get them all done!" No: people have to prioritize and focus in order to be productive, and in order to do that, they have to be taught how. One widely used technique is to make a list of things that need to be done, sort them by priority, and then switch off email and other interruptions for 30–60 minutes and complete one of those tasks. If any task on a to-do list is more than an hour long, break it down into smaller pieces and prioritize those separately.

The most important part of this is switching off interruptions. Despite what many people want to believe, human beings are not good at multi-tasking. What we can become good at is **automaticity**, which is the ability to do something routine in the background while doing something else [Mill2016a]. Most of us can talk while chopping onions, or drink coffee while reading; with practice, we can also take notes while listening, but we can't study effectively, program, or do other mentally challenging tasks while paying attention to something else—we only think we can.

The point of organizing and preparing is to get into the most productive mental state possible. Psychologists call it **flow** [Csik2008]; athletes call it "being in the zone," and musicians talk about losing themselves in what they're playing. Whatever name you use, people produce much more per unit of time in this state than normal. The bad news is that it takes roughly 10 minutes to get back into a state of flow after an interruption, no matter how short the interruption was. This means that if you are interrupted half a dozen times per hour, you are *never* at your productive peak.

How Did He Know?

In his 1961 short story "Harrison Bergeron,[5]*" Kurt Vonnegut described a future in which everyone is forced to be equal. Good-looking people have to wear masks, athletic people have to wear weights—and intelligent people are forced to carry around radios that interrupt their thoughts at random intervals. I sometimes wonder if—oh, hang on, my phone just— sorry, what were we talking about?*

5.3 PEER ASSESSMENT

Asking people on a team to rate their peers is a common practice in industry. [Sond2012] surveyed the literature on peer assessment, distinguishing between grading and reviewing. They found that peer assessment increased the amount, diversity, and timeliness of feedback, helped learners exercise higher-level thinking, encouraged reflective practice, and supported development of social skills. The concerns were predictable: validity and reliability, motivation and procrastination, trolls, collusion, and plagiarism.

However, the evidence shows that these concerns aren't significant in most classes. For example, [Kauf2000] compared confidential peer ratings and grades on several axes for two undergraduate engineering courses and found that self-rating and peer ratings statistically agreed, that collusion wasn't significant (i.e., people didn't just give all their peers high grades), that learners didn't inflate their self-ratings, and crucially, that ratings were not biased by gender or race.

One way to implement peer assessment is **contributing student pedagogy**, in which learners produce artifacts to contribute to others' learning. This can be developing a short lesson and sharing it with the class, adding to a question bank, or writing up notes from a particular lecture for in-class publication. For example, [Fran2018] found that learners who made short videos to teach concepts to their peers had a significant increase in their own learning compared to those who only studied the material or viewed the videos. I have found that asking learners to share one bug and its fix with the class every day helps their analytic abilities and reduces impostor syndrome.

Another approach is **calibrated peer review**, in which a learner reviews one or more examples using a rubric and compares their evaluation against the teacher's review of the same work [Kulk2013]. Once learners' evaluations are close enough to the teacher's, they start evaluating their peers' actual work. If several peers' assessments are combined, this can be as accurate as assessment by teachers [Pare2008].

Like everything else, assessment is aided by rubrics. The evaluation form in Section F.2 shows a sample to get you started. To use it, rank yourself and each of your teammates, then calculate and compare scores. Large disparities usually indicate a need for a longer conversation.

---

[5]https://en.wikipedia.org/wiki/Harrison_Bergeron

5.4   EXERCISES

LEARNING STRATEGIES (INDIVIDUAL/20)

1. Which of the six learning strategies do you regularly use? Which ones do you not?
2. Write down three general concepts that you want your learners to master and give two specific examples of each (concrete examples practice). For each of those concepts, work backward from one of your examples to explain how the concept explains it (elaboration).

CONNECTING IDEAS (PAIRS/5)

This exercise is an example of using elaboration to improve retention. Pick a partner have each person independently choose an idea, then announce your ideas and try to find a four-link chain that leads from one to the other. For example, if the two ideas are "Saskatchewan" and "statistics," the links might be:

- Saskatchewan is a province of Canada;
- Canada is a country;
- countries have governments;
- governments use statistics to analyze public opinion.

CONVERGENT EVOLUTION (PAIRS/15)

One practice that wasn't covered above is **guided notes**, which are notes prepared by the teacher that cue learners to respond to key information in a lecture or discussion. The cues can be blank spaces where learners add information, asterisks next to terms learners should define, and so on.

Create 2–4 guided note cards for a lesson you have recently taught or are going to teach. Swap cards with your partner: how easy is it to understand what is being asked for? How long would it take to fill in the prompts? How well does this work for programming examples?

CHANGING MINDS (PAIRS/10)

[Kirs2013] argues that myths about digital natives, learning styles, and self-educators are all reflections of the mistaken belief that learners know what is best for them, and cautions that we may be in a downward spiral in which every attempt by education researchers to rebut these myths confirms their opponents' belief that learning science is pseudo-science. Pick one thing you have learned about learning so far in this book that surprised you or contradicted something you previously believed and practice explaining it to a partner in 1–2 minutes. How convincing are you?

FLASH CARDS (INDIVIDUAL/15)

Use sticky notes or anything else you have at hand to make up half a dozen flash cards for a topic you have recently taught or learned. Trade with a partner and see how long it takes each of you to achieve 100% perfect recall. Set the cards aside when you are done, then come back after half an hour and see what your recall rate is.

USING ADEPT (WHOLE CLASS/15)

Pick something you have recently taught or been taught and outline a short lesson that uses the five-step ADEPT method to introduce it.

THE COST OF MULTI-TASKING (PAIRS/10)

The Learning Scientists blog[6] describes a simple experiment you can do with only a stopwatch to demonstrate the mental cost of multi-tasking. Working in pairs, measure how long it takes each person to do each of these three tasks:

- Count from 1 to 26 twice.
- Recite the alphabet from A to Z twice.
- Interleave the numbers and letters, i.e., say, "1, A, 2, B, ..." and so on.

Have each pair report their numbers. Without specific practice, the third task always takes significantly longer than either of the component tasks.

MYTHS IN COMPUTING EDUCATION (WHOLE CLASS/20)

[Guzd2015b] presents a list of the top 10 mistaken beliefs about computing education, which includes:

1. The lack of women in Computer Science is just like all the other STEM fields.
2. To get more women in CS, we need more female CS faculty.
3. Student evaluations are the best way to evaluate teaching.
4. Good teachers personalize education for students' learning styles.
5. A good CS teacher should model good software development practice because their job is to produce excellent software engineers.
6. Some people are just naturally better programmers than others.

Have everyone vote +1 (agree), -1 (disagree), or 0 (not sure) for each point, then read the full explanations in the original article[7] and vote again. Which ones did people change their minds on? Which ones do they still believe are true, and why?

---

[6]http://www.learningscientists.org/blog/2017/7/28-1

[7]https://cacm.acm.org/blogs/blog-cacm/189498-top-10-myths-about-teaching-computer-science/fulltext

CALIBRATED PEER REVIEW (PAIRS/20)

1. Create a 5–10 point rubric with entries like "good variable names," "no redundant code," and "properly-nested control flow" for grading the kind of programs you would like your learners to write.
2. Choose or create a small program that contains 3–4 violations of these entries.
3. Grade the program according to your rubric.
4. Have a partner grade the same program with the same rubric. What do they accept that you did not? What do they critique that you did not?

REVIEW

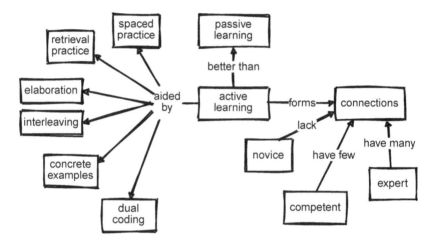

Figure 5.1: Concepts: Active learning

6 A Lesson Design Process

Most people design lessons like this:

1. Someone asks you to teach something you barely know or haven't thought about in years.
2. You start writing slides to explain what you know about the subject.
3. After 2 or 3 weeks, you make up an assignment based on what you've taught so far.
4. You repeat step 3 several times.
5. You stay awake into the wee hours of the morning to create a final exam and promise yourself that you'll be more organized next time.

A more effective method is similar in spirit to a programming practice called **test-driven development** (TDD). Programmers who use TDD don't write software and then test that it is working correctly. Instead, they write the tests first, then write just enough new software to make those tests pass.

TDD works because writing tests forces programmers to be precise about what they're trying to accomplish and what "done" looks like. TDD also prevents endless polishing: when the tests pass, you stop coding. Finally, it reduces the risk of confirmation bias: someone who hasn't yet written a piece of software will be more objective than someone who has just put in several hours of hard work and really, really wants to be done.

A similar method called **backward design** works very well for lesson design. This method was developed independently in [Wigg2005; Bigg2011; Fink2013] and is summarized in [McTi2013]. In simplified form, its steps are:

1. Create or recycle learner personas (discussed in the next section) to figure out who you are trying to help and what will appeal to them.
2. Brainstorm to get a rough idea of what you want to cover, how you're going to do it, what problems or misconceptions you expect to encounter, what's *not* going to be included, and so on. Drawing concept maps can help a lot at this stage (Section 3.1).
3. Create a summative assessment (Section 2.1) to define your overall goal. This can be the final exam for a course or the capstone project for a one-day workshop; regardless of its form or size, it shows how far you hope to get more clearly than a point-form list of objectives.
4. Create formative assessments that will give people a chance to practice the things they're learning. These will also tell you (and them) whether they're making progress and where they need to focus their attention. The best way to do this is to itemize the knowledge and skills used in the summative assessment you developed in the previous step and then create at least one formative assessment for each.

5. Order the formative assessments to create a course outline based on their complexity, their dependencies, and how well topics will motivate your learners (Section 10.1).
6. Write material to get learners from one formative assessment to the next. Each hour of instruction should consist of three to five such episodes.
7. Write a summary description of the course to help its intended audience find it and figure out whether it's right for them.

This method helps keep teaching focused on its objectives. It also ensures that learners don't face anything at the end of the course that they are not prepared for.

Perverse Incentives

Backward design is not *the same thing as* **teaching to the test**. *When using backward design, teachers set goals to aid in lesson design; they may never actually give the final exam that they wrote. In many school systems, on the other hand, an external authority defines assessment criteria for all learners, regardless of their individual situations. The outcomes of those summative assessments directly affect the teachers' pay and promotion, which means teachers have an incentive to focus on having learners pass tests rather than on helping them learn.*

[Gree2014] argues that focusing on testing and measurement appeals to those with the power to set the tests, but is unlikely to improve outcomes unless it is coupled with support for teachers to make improvements based on test outcomes. The latter is often missing because large organizations usually value uniformity over productivity [Scot1998].

Reverse design is described as a sequence, but it's almost never done that way. We may, for example, change our mind about what we want to teach based on something that occurs to us while we're writing an MCQ, or re-assess who we're trying to help once we have a lesson outline. However, the notes we leave behind should present things in the order described above so that whoever has to use or maintain the lesson after us can retrace our thinking [Parn1986].

6.1 LEARNER PERSONAS

The first step in the reverse design is figuring out who your audience is. One way to do this is to write two or three **learner personas** like those in Section 1.1. This technique is borrowed from user experience designers, who create short profiles of typical users to help them think about their audience.

A learner persona consists of:

1. the person's general background;
2. what they already know;
3. what they want to do; and
4. any special needs they have.

The personas in Section 1.1 have the four points listed above, along with a short summary of how this book will help them. A learner persona for a volunteer group that runs weekend Python workshops might be:

1. Jorge just moved from Costa Rica to Canada to study agricultural engineering. He has joined the college soccer team and is looking forward to learning how to play ice hockey.
2. Other than using Excel, Word, and the Internet, Jorge's most significant previous experience with computers is helping his sister build a WordPress site for the family business back home.
3. Jorge wants to measure properties of soil from nearby farms using a handheld device that sends data to his computer. Right now he has to open each data file in Excel, delete the first and last column, and calculate some statistics on what's left. He has to collect at least 600 measurements in the next few months, and really doesn't want to have to do these steps by hand for each one.
4. Jorge can read English well, but sometimes struggles to keep up with spoken conversation that involves a lot of jargon.

Rather than writing new personas for every lesson or course, teachers usually create and share half a dozen that cover everyone they are likely to teach, then pick a few from that set to describe the audience for particular material. Personas that are used this way become a convenient shorthand for design issues: when speaking with each other, teachers can say, "Would Jorge understand why we're doing this?" or, "What installation problems would Jorge face?"

Their Goals, Not Yours
Personas should always describe what the learner wants to do rather than what you think they actually need. Ask yourself what they are searching for online; it probably won't include jargon that they don't yet know, so part of what you have to do as an instructional designer is figure out how to make your lesson findable.

6.2 LEARNING OBJECTIVES

Formative and summative assessments help teachers figure out what they're going to teach, but in order to communicate that to learners and other teachers, a course description should also have **learning objectives**. These help ensure that everyone has the same understanding of what a lesson is supposed to accomplish. For example, a statement like "understand Git" could mean any of the following:

- Learners can describe three ways in which version control systems like Git are better than file-sharing tools like Dropbox and two ways in which they are worse.
- Learners can commit a changed file to a Git repository using a desktop GUI tool.
- Learners can explain what a detached HEAD is and recover from it using command-line operations.

Objectives vs. Outcomes

*A learning objective is what a lesson strives to achieve. A **learning outcome** is what it actually achieves, i.e., what learners actually take away. The role of summative assessment is therefore to compare learning outcomes with learning objectives.*

A learning objective describes how the learner will demonstrate what they have learned once they have successfully completed a lesson. More specifically, it has a *measurable or verifiable verb* that states what the learner will do and specifies the *criteria for acceptable performance*. Writing these may initially seem restrictive, but they will make you, your fellow teachers, and your learners happier in the long run: you will end up with clear guidelines for both your teaching and assessment, and your learners will appreciate having clear expectations.

One way to understand what makes for a good learning objective is to see how a poor one can be improved:

- *The learner will be given opportunities to learn good programming practices.*
  This describes the lesson's content, not the attributes of successful learners.

- *The learner will have a better appreciation for good programming practices.*
  This doesn't start with an active verb or define the level of learning, and the subject of learning has no context and is not specific.

- *The learner will understand how to program in R.*
  While this starts with an active verb, it doesn't define the level of learning and the subject of learning is still too vague for assessment.

- *The learner will write one-page data analysis scripts to read, filter, and summarize tabular data using R.*
  This starts with an active verb, defines the level of learning, and provides context to ensure that outcomes can be assessed.

When it comes to choosing verbs, many teachers use **Bloom's Taxonomy**. First published in 1956 and updated at the turn of the century [Ande2001], it is a widely used framework for discussing levels of understanding. Its most recent form has six categories; the list below gives a few of the verbs typically used in learning objectives written for each:

Remembering: Exhibit memory of previously learned material by recalling facts, terms, basic concepts, and answers. *(recognize, list, describe, name, find)*

Understanding: Demonstrate understanding of facts and ideas by organizing, comparing, translating, interpreting, giving descriptions, and stating main ideas. *(interpret, summarize, paraphrase, classify, explain)*

Applying: Solve new problems by applying acquired knowledge, facts, techniques and rules in a different way. *(build, identify, use, plan, select)*

Analyzing: Examine and break information into parts by identifying motives or causes; make inferences and find evidence to support generalizations. *(compare, contrast, simplify)*

Evaluating: Present and defend opinions by making judgments about information, validity of ideas, or quality of work based on a set of criteria. *(check, choose, critique, prove, rate)*

Creating: Compile information together in a different way by combining elements in a new pattern or proposing alternative solutions. *(design, construct, improve, adapt, maximize, solve)*

Bloom's Taxonomy appears in almost every textbook on education, but [Masa2018] found that even experienced educators have trouble agreeing on how to classify specific things. The verbs are still useful, though, as is the notion of building understanding in steps: as Daniel Willingham has said, people can't think without something to think about [Will2010], and this taxonomy can help teachers ensure that learners have those somethings when they need them.

Another way to think about learning objectives comes from [Fink2013], which defines learning in terms of the change it is meant to produce in the learner. **Fink's Taxonomy** also has six categories, but unlike Bloom's they are complementary rather than hierarchical:

Foundational Knowledge: understanding and remembering information and ideas. *(remember, understand, identify)*

Application: skills, critical thinking, managing projects. *(use, solve, calculate, create)*

Integration: connecting ideas, learning experiences, and real life. *(connect, relate, compare)*

Human Dimension: learning about oneself and others. *(come to see themselves as, understand others in terms of, decide to become)*

Caring: developing new feelings, interests, and values. *(get excited about, be ready to, value)*

Learning How to Learn: becoming a better learner. *(identify source of information for, frame useful questions about)*

A set of learning objectives based on this taxonomy for an introductory course on HTML and CSS might be:

- Explain what CSS properties are and how CSS selectors work.
- Style a web page using common tags and CSS properties.
- Compare and contrast writing HTML and CSS to writing with desktop publishing tools.
- Identify and correct issues in sample web pages that would make them difficult for the visually impaired to interact with.

- Describe features of favorite web sites whose design particularly appeals to you and explain why.
- Describe your two favorite online sources of information about CSS and explain what you like about them.

6.3 MAINTAINABILITY

Once a lesson has been created someone needs to maintain it, and doing that is a lot easier if it has been built in a maintainable way. But what exactly does "maintainable" mean? The short answer is that a lesson is maintainable if it's cheaper to update it than to replace it. This equation depends on four factors:

How well documented the course's design is. If the person doing maintenance doesn't know (or doesn't remember) what the lesson is supposed to accomplish or why topics are introduced in a particular order, it will take them more time to update it. One reason to use reverse design is to capture decisions about why each course is the way it is.

How easy it is for collaborators to collaborate technically. Teachers usually share material by mailing PowerPoint files to each other or by putting them in a shared drive. Collaborative writing tools like Google Docs[1] and wikis are a big improvement, as they allow many people to update the same document and comment on other people's updates. The version control systems used by programmers, such as GitHub[2], are another approach. They let any number of people work independently and then merge their changes in a controlled, reviewable way. Unfortunately, version control systems have a steep learning curve and don't handle common office document formats.

How willing people are to collaborate. The tools needed to build a Wikipedia for lessons have been around for twenty years, but most teachers still don't write and share lessons the way that they write and share encyclopedia entries.

How useful sharing actually is. The **Reusability Paradox** states that the more reusable a learning object is, the less pedagogically effective it is [Wile2002]. The reason is that a good lesson resembles a novel more than it does a program: its parts are tightly coupled rather than independent black boxes. Direct re-use may therefore be the wrong goal for lessons; we might get further by trying to make them easier to remix.

If the Reusability Paradox is true, collaboration will be more likely if the things being collaborated on are small. This fits well with Mike Caulfield's theory of choral explanations[3], which argues that sites like Stack Overflow[4] succeed because they

---

[1]http://docs.google.com

[2]http://github.com

[3]https://hapgood.us/2016/05/13/choral-explanations/

[4]https://stackoverflow.com/

provide a chorus of answers for every question, each of which is most suitable for a slightly different questioner. If this is right, the lessons of tomorrow may be guided tours of community-curated Q&A repositories designed for learners at widely different levels.

6.4   EXERCISES

CREATE LEARNER PERSONAS (SMALL GROUPS/30)

Working in small groups, create a 4-point persona that describes one of your typical learners.

CLASSIFY LEARNING OBJECTIVES (PAIRS/10)

Look at the example learning objectives for an introductory course on HTML and CSS in Section 6.2 and classify each according to Bloom's Taxonomy. Compare your answers with those of your partner. Where did you agree and disagree?

WRITE LEARNING OBJECTIVES (PAIRS/20)

Write one or more learning objectives for something you currently teach or plan to teach using Bloom's Taxonomy. Working with a partner, critique and improve the objectives. Does each one have a verifiable verb and clearly state criteria for acceptable performance?

WRITE MORE LEARNING OBJECTIVES (PAIRS/20)

Write one or more learning objectives for something you currently teach or plan to teach using Fink's Taxonomy. Working with a partner, critique and improve the objectives.

HELP ME DO IT BY MYSELF (SMALL GROUPS/15)

The educational theorist Lev Vygotsky introduced the notion of a **Zone of Proximal Development** (ZPD), which includes the problems that people cannot yet solve on their own but are able to solve with help from a mentor. These are the problems that are most fruitful to tackle next, as they are out of reach but attainable.

Working in small groups, choose one learner persona you have developed and describe two or three problems that are in that learner's ZPD.

BUILDING LESSONS BY SUBTRACTING COMPLEXITY (INDIVIDUAL/20)

One way to build a programming lesson is to write the program you want learners to finish with, then remove the most complex part that you want them to write and make it the last exercise. You can then remove the next most complex part you want them to write and make it the penultimate exercise, and so on. Anything that's left after you have pulled out the exercises, such as loading libraries or reading data, becomes the starter code that you give them.

Take a program or web page that you want your learners to be able to create and work backward to break it into digestible parts. How many are there? What key idea is introduced by each one?

INESSENTIAL WEIRDNESS (INDIVIDUAL/15)

Betsy Leondar-Wright coined the phrase "inessential weirdness[5]" to describe things groups do that aren't really necessary, but which alienate people who aren't yet members of that group. Sumana Harihareswara later used this notion as the basis for a talk on inessential weirdnesses in open source software[6], which includes things like using command-line tools with cryptic names. Take a few minutes to read these articles, then make a list of inessential weirdnesses you think your learners might encounter when you first teach them. How many of these can you avoid?

PETE (INDIVIDUAL/15)

One pattern that works well for programming lessons is PETE: introduce the **P**roblem, work through an **E**xample, explain the **T**heory, and then **E**laborate on a second example so that learners can see what is specific to each case and what applies to all cases. Pick something you have recently taught or been taught and outline a short lesson for it that follows these five steps.

PRIMM (INDIVIDUAL/15)

Another lesson pattern is PRIMM [Sent2019]: **P**redict a program's behavior or output, **R**un it to see what it actually does, **I**nvestigate why it does that by stepping through it in a debugger or drawing the flow of control, **M**odify it (or its inputs), and then **M**ake something similar from scratch. Pick something you have recently taught or been taught and outline a short lesson for it that follows these five steps.

CONCRETE-REPRESENTATIONAL-ABSTRACT (PAIRS/15)

Concrete-Representational-Abstract[7] (CRA) is an approach to introducing new ideas that is used primarily with younger learners: physically manipulate a **C**oncrete object, **R**epresent the object with an image, then perform the same operations using numbers, symbols, or some other **A**bstraction.

1. Write each of the numbers 2, 7, 5, 10, 6 on a sticky note.
2. Simulate a loop that finds the largest value by looking at each in turn (concrete).

[5]http://www.classmatters.org/2006_07/its-not-them.php

[6]https://www.harihareswara.net/sumana/2016/05/21/0

[7]https://makingeducationfun.wordpress.com/2012/04/29/concrete-representational-abstract-cra/

3. Sketch a diagram of the process you used, labeling each step (representational).
4. Write instructions that someone else could follow to go through the same steps (abstract).

Compare your representational and abstract materials with your partner's.

EVALUATING A LESSON REPOSITORY (SMALL GROUPS/10)

[Leak2017] explores why computer science teachers don't use lesson sharing sites and recommends ways to make them more appealing:

1. The landing page should allow site visitors to identify their background and their interests in visiting the site. Sites should ask two questions: "What is your current role?" and "What course and grade level are you interested in?"
2. Sites should display all learning resources in the context of the full course so that potential users can understand their intended context of use.
3. Many teachers have concerns about having their (lack of) knowledge judged by their peers if they post to sites' discussion forums. These forums should therefore allow anonymous posting.

In small groups, discuss whether these three features would be enough to convince you to use a lesson sharing site, and if not, what would.

REVIEW

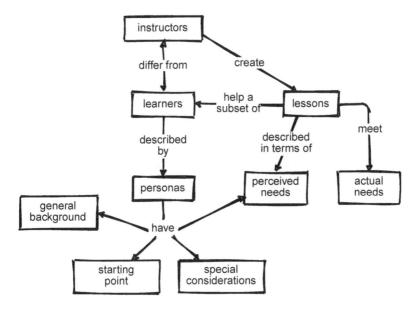

Figure 6.1: Concepts: Learner personas

7 Pedagogical Content Knowledge

Every teacher needs three things:

content knowledge such as how to program;

general pedagogical knowledge such as an understanding of the psychology of learning; and

pedagogical content knowledge (PCK), which is the domain-specific knowledge of how to teach a particular concept to a particular audience. In computing, PCK includes things like what examples to use when teaching how parameters are passed to a function or what misconceptions about nesting HTML tags are most common.

We can add technical knowledge to this mix [Koeh2013], but that doesn't change the key point: it isn't enough to know the subject and how to teach—you have to know how to teach that particular subject [Maye2004]. This chapter therefore summarizes some results from computing education research that will add to your store of PCK.

As with all research, some caution is required when interpreting results:

Theories change as more data becomes available. Computing education research (CER) is a young discipline: the American Society for Engineering Education was founded in 1893 and the National Council of Teachers of Mathematics in 1920, but the Computer Science Teachers Association didn't exist until 2005. While a steady stream of new insights are reported at conferences like SIGCSE[1], ITiCSE[2], and ICER[3], we simply don't know as much about learning to program as we do about learning to read, play a sport, or do basic arithmetic.

Most of these studies' subjects are WEIRD: they are from Western, Education, Industrialized, Rich, and Democratic societies [Henr2010]. What's more, they are also mostly young and in institutional classrooms, since that's the population most researchers have easiest access to. We know much less about adults, members of marginalized groups, learners in free-range settings, or **end-user programmers**, even though there are far more of them.

---

[1] http://sigcse.org/

[2] http://iticse.acm.org/

[3] https://icer.hosting.acm.org

If this was an academic treatise, I would therefore preface most claims with qualifiers like, "Some research may seem to indicate that..." But since actual teachers in actual classrooms have to make decisions regardless of whether research has clear answers yet or not, this chapter presents actionable best guesses rather than nuanced perhapses.

Jargon

*Like any specialty, CER has jargon. **CS1** refers to an introductory semester-long course in which learners meet variables, loops, and functions for the first time, while **CS2** refers to a second course that covers basic data structures like stacks and queues, and **CS0** refers to an introductory course for people without any prior experience who aren't intending to continue with computing right away. Full definitions for these terms can be found in the ACM Curriculum Guidelines[4].*

7.1   WHAT ARE WE TEACHING THEM NOW?

Very little is known about what coding bootcamps and other free-range initiatives teach, in part because many are reluctant to share their curriculum. We know more about what is taught by institutions [Luxt2017]:

Topic	%	Topic	%
Programming Process	87%	Data Types	23%
Abstract Programming Thinking	63%	Input/Output	17%
Data Structures	40%	Libraries	15%
Object-Oriented Concepts	36%	Variables & Assignment	14%
Control Structures	33%	Recursion	10%
Operations & Functions	26%	Pointers & Memory Management	5%

High-level topic labels like these can hide a multitude of sins. A more tangible result comes from [Rich2017], which reviewed a hundred articles to find learning trajectories for computing classes in elementary and middle schools. Their results for sequencing, repetition, and conditionals are essentially collective concept maps that combine and rationalize the implicit and explicit thinking of many different educators (Figure 7.1).

---

[4]https://www.acm.org/education/curricula-recommendations

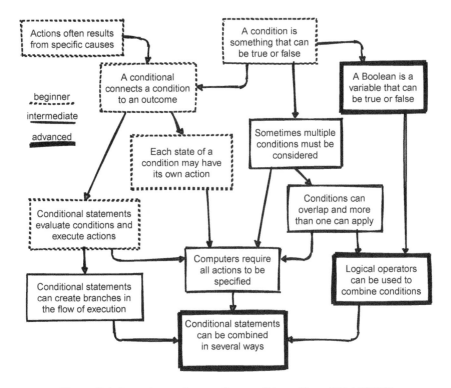

Figure 7.1: Learning trajectory for conditions (from [Rich2017])

7.2 HOW MUCH ARE THEY LEARNING?

There can be a world of difference between what teachers teach and how much learners learn. To explore the latter, we must use other measures or do direct studies. Taking the former approach, roughly two-thirds of post-secondary students pass their first computing course, with some variations depending on class size and so on, but with no significant differences over time or based on language [Benn2007a; Wats2014].

How does prior experience affect these results? To find out, [Wilc2018] compared the performance and confidence of novices with and without prior programming experience in CS1 and CS2 (see below). They found that novices with prior experience outscored novices without by 10% in CS1, but those differences disappeared by the end of CS2. They also found that women with prior exposure outperformed their male peers in all areas, but were consistently less confident in their abilities (Section 10.4).

As for direct studies of how much novices learn, [McCr2001] presented a multi-site international study that was later replicated by [Utti2013]. According to the first study, "...the disappointing results suggest that many students do not know how to program at the conclusion of their introductory courses." More specifically, "For a combined sample of 216 students from four universities, the average score was 22.89 out of 110 points on the general evaluation criteria developed for this study." This result may say as much about teachers' expectations as it does about student ability, but either way, our first recommendation is to *measure and track results* in ways that can be compared over time so that you can tell if your lessons are becoming more or less effective.

7.3   WHAT MISCONCEPTIONS DO NOVICES HAVE?

Chapter 2 explained why clearing up novices' misconceptions is just as important as teaching them strategies for solving problems. The biggest misconception novices have—sometimes called the "superbug" in coding—is the belief that computers understand what people mean in the way that another human being would [Pea1986]. Our second recommendation is therefore to *teach novices that computers don't understand programs*, i.e., that calling a variable "cost" doesn't guarantee that its value is actually a cost.

[Sorv2018] presents over forty other misconceptions that teachers can also try to clear up, many of which are also discussed in [Qian2017]'s survey. One is the belief that variables in programs work the same way they do in spreadsheets, i.e., that after executing:

```
grade = 65
total = grade + 10
grade = 80
print(total)
```

the value of `total` will be 90 rather than 75 [Kohn2017]. This is an example of the way in which novices construct a plausible-but-wrong mental model by making analogies; other misconceptions include:

- A variable holds the history of the values it has been assigned, i.e., it remembers what its value used to be.
- Two objects with the same value for a `name` or `id` attribute are guaranteed to be the same object.
- Functions are executed as they are defined, or are executed in the order in which they are defined.
- A `while` loop's condition is constantly evaluated, and the loop stops as soon as it becomes false. Conversely, the conditions in `if` statements are also constantly evaluated, and their statements are executed as soon as the condition becomes true regardless of where the flow of control is at the time.
- Assignment moves values, i.e., after a = b, the variable b is empty.

7.4   WHAT MISTAKES DO NOVICES MAKE?

The mistakes novices make can tell us what to prioritize in our teaching, but it turns out that most teachers don't know how common different kinds of mistakes actually are. The largest study of this is [Brow2017], which found that mismatched quotes and parentheses are the most common type of errors in novice Java programs, but also the easiest to fix, while some mistakes (like putting the condition of an if in {} instead of ()) are most often made only once. Unsurprisingly, mistakes that produce compiler errors are fixed much faster than ones that don't. Some mistakes, however, are made many times, like invoking methods with the wrong arguments (e.g., passing a string instead of an integer).

Not Right vs. Not Done
One difficulty in research like this is distinguishing mistakes from work in progress. For example, an empty if statement or a method that is defined but not yet used may be a sign of incomplete code rather than an error.

[Brow2017] also compared the mistakes novices actually make with what their teachers thought they made. They found that, "...educators formed only a weak consensus about which mistakes are most frequent, that their rankings bore only a moderate correspondence to the students in the...data, and that educators' experience had no effect on this level of agreement." For example, mistaking = (assignment) and == (equality) wasn't nearly as common as most teachers believed.

Not Just for Code
[Park2015] collected data from an online HTML editor during an introductory web development course and found that nearly all learners made syntax errors that remained unresolved weeks into the course. 20% of these errors related to the relatively complex rules that dictate when *it is valid for HTML elements to be nested in one another, but 35% related to the simpler tag syntax determining* how *HTML elements are nested. The tendency of many teachers to say, "But the rules are simple," is a good example of expert blind spot discussed in Chapter 3...*

7.5 HOW DO NOVICES PROGRAM?

[Solo1984; Solo1986] pioneered the exploration of novice and expert programming strategies. The key finding is that experts know both "what" and "how," i.e., they understand what to put into programs *and* they have a set of program patterns or plans to guide their construction. Novices lack both, but most teachers focus solely on the former, even though bugs are often caused by not having a strategy for solving the problem rather than to lack of knowledge about the language. Recent work has shown the effectiveness of teaching four distinct skills in a specific order [Xie2019]:

	semantics of code	**templates related to goals**
reading	1. read code and predict behavior	3. recognize templates and their uses
writing	2. write correct syntax	4. use templates to meet goals

Our next recommendations are therefore to *have learners read code, then modify it, then write it*, and to *introduce common patterns explicitly and have learners practice using them.* [Mull2007b] is just one of many studies proving the benefits of teaching common patterns explicitly, and decomposing problems into patterns creates natural opportunities for creating and labeling subgoals [Marg2012; Marg2016].

7.6   HOW DO NOVICES DEBUG?

A decade ago, [McCa2008] wrote, "It is surprising how little page space is devoted to bugs and debugging in most introductory programming textbooks." Little has changed since: there are hundreds of books on compilers and operating systems, but only a handful about debugging, and I have never seen an undergraduate course devoted to the subject.

[List2004; List2009] found that many novices struggled to predict the output of short pieces of code and to select the correct completion of the code from a set of possibilities when told what it was supposed to do. More recently, [Harr2018] found that the gap between being able to trace code and being able to write it has largely closed by CS2, but that novices who still have a gap (in either direction) are likely to do poorly.

Our fifth recommendation is therefore to *explicitly teach novices how to debug.* [Fitz2008; Murp2008] found that good debuggers were good programmers, but not all good programmers were good at debugging. Those who were used a symbolic debugger to step through their programs, traced execution by hand, wrote tests, and re-read the spec frequently, which are all teachable habits. However, tracing execution step by step was sometimes used ineffectively: for example, a novice might put the same `print` statement in both parts of an `if-else`. Novices would also comment out lines that were actually correct as they tried to isolate a problem; teachers can make both of these mistakes deliberately, point them out, and correct them to help novices get past them.

Teaching novices how to debug can also help make classes easier to manage. [Alqa2017] found that learners with more experience solved debugging problems significantly faster, but times varied widely: 4–10 minutes was a typical range for individual exercises, which means that some learners need 2–3 times longer than others to get through the same exercises. Teaching the slower learners what the faster ones are doing will help make the group's overall progress more uniform.

Debugging depends on being able to read code, which multiple studies have shown is the single most effective way to find bugs [Basi1987; Keme2009; Bacc2013]. The code quality rubric developed in [Steg2014; Steg2016a] is a good checklist of things to look for, though it is best presented in chunks rather than all at once.

Having learners read code and summarize its behavior is a good exercise (Section 5.1), but often takes too long to be practical in class. Having them predict a program's output just before it is run, on the other hand, helps reinforce learning (Section 9.11) and also gives them a natural moment to ask "what if" questions. Teachers or learners can also trace changes to variables as they go along, which [Cunn2017] found was effective (Section 12.2).

7.7 WHAT ABOUT TESTING?

Novice programmers seem just as reluctant to test software as professionals. There's no doubt that doing it is valuable—[Cart2017] found that high-performing novices spent a lot of time testing, while low performers spent much more time working on code with errors—and many teachers require learners to write tests for assignments. But how well do they do this? One answer comes from [Bria2015], which scored learners' programs by how many teacher-provided test cases those programs passed, and conversely scores test cases written by learners according to how many deliberately-seeded bugs they caught. They found that novices' tests often have low coverage (i.e., they don't test most of the code) and that they often test many things at once, which makes it hard to pinpoint the causes of errors.

Another answer comes from [Edwa2014b], which looked at all of the bugs in all novices' code submissions combined and identified those detected by the novices' test suite. They found that novices' tests only detected an average of 13.6% of the faults present in the entire program population. What's more, 90% of the novices' tests were very similar, which indicates that novices mostly write tests to confirm that code is doing what it's supposed to rather than to find cases where it isn't.

One approach to teaching better testing practices is to define a programming problem by providing a set of tests to be passed rather than through a written description (Section 12.1). Before doing this, though, take a moment to look at how many tests you've written for your own code recently, and then decide whether you're teaching what you believe people should do, or what they (and you) actually do.

7.8 DO LANGUAGES MATTER?

The short answer is "yes": novices learn to program faster and learn more using blocks-based tools like Scratch (Figure 7.2) [Wein2017]. One reason is that blocks-based systems reduce cognitive load by eliminating the possibility of syntax errors. Another is that block interfaces encourage exploration in a way that text does not: like all good tools, Scratch can be learned accidentally [Malo2010].

But what happens *after* blocks? [Chen2018] found that learners whose first programming language was graphical had higher grades in introductory programming courses than learners whose first language was textual when the languages were introduced in or before early adolescent years. Our sixth recommendation is therefore to *start children and teens with blocks-based interfaces* before moving to text-based systems. The age qualification is there because Scratch deliberately looks like it's meant for younger users, and it can still be hard to convince adults to take it seriously.

Scratch has probably been studied more than any other programming tool. One example is [Aiva2016], which analyzed over 250,000 Scratch projects and found (among other things) that about 28% of projects have some blocks that are never called or triggered. As in the earlier aside about incomplete versus incorrect Java programs, the authors hypothesize that users may be using these blocks as a scratchpad to keep track of bits of code they don't (yet) want to throw away. Another example

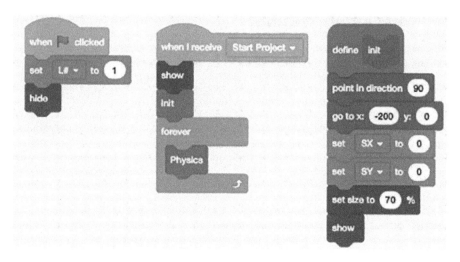

Figure 7.2: Scratch

is [Grov2017; Mlad2017], which studied novices learning about loops in Scratch, Logo, and Python. They found that misconceptions about loops are minimized when using a block-based language rather than a text-based language. What's more, as tasks become more complex (such as using nested loops) the differences become larger.

Harder Than Necessary

The creators of programming language make those languages harder to learn by not doing basic usability testing. For example, [Stef2013] found that, "...the three most common words for looping in computer science, for, while, *and* foreach, *were rated as the three most unintuitive choices by non-programmers." Their work shows that C-style syntax (as used in Java and Perl) is just as hard for novices to learn as a randomly designed syntax, but that the syntax of languages such as Python and Ruby is significantly easier to learn, and the syntax of a language whose features are tested before being added to the language is easier still. [Stef2017] is a useful brief summary of what we actually know about designing programming languages and why we believe it's true, while [Guzd2016] lays out five principles that programming languages for learners should follow.*

OBJECT-ORIENTED PROGRAMMING

Objects and classes are power tools for experienced programmers, and many educators advocate an **objects first** approach to teaching programming (though they sometimes disagree on exactly what that means [Benn2007b]). [Sorv2014] describes and motivates this approach, and [Koll2015] describes three generations of tools designed to support novice programming in object-oriented environments.

Introducing objects early has a few challenges. [Mill2016b] found that most novices using Python struggled to understand `self` (which refers to the current object): they omitted it in method definitions, failed to use it when referencing object attributes, or both. [Rago2017] found something similar in high school students, and also found that high school teachers often weren't clear on the concept either. On balance, we recommend that teachers *start with functions rather than objects*, i.e., that learners not be taught how to define classes until they understand basic control structures and data types.

TYPE DECLARATIONS

Programmers have argued for decades about whether variables' data types should have to be declared or not, usually based on their personal experience as professionals rather than on any kind of data. [Endr2014; Fisc2015] found that requiring novices to declare variable types does add some complexity to programs, but it pays off fairly quickly by acting as documentation for a method's use—in particular, by forestalling questions about what's available and how to use it.

VARIABLE NAMING

[Kern1999] wrote, "Programmers are often encouraged to use long variable names regardless of context. This is a mistake: clarity is often achieved through brevity." Lots of programmers believe this, but [Hofm2017] found that using full words in variable names led to an average of 19% faster comprehension compared to letters and abbreviations. In contrast, [Beni2017] found that using single-letter variable names didn't affect novices' ability to modify code. This may be because their programs are shorter than professionals' or because some single-letter variable names have implicit types and meanings. For example, most programmers assume that i, j, and n are integers and that s is a string, while x, y, and z are either floating-point numbers or integers more or less equally.

How important is this? [Bink2012] reported that reading and understanding code is fundamentally different from reading prose: "...the more formal structure and syntax of source code allows programmers to assimilate and comprehend parts of the code quite rapidly independent of style. In particular...beacons and program plans play a large role in comprehension." It also found that experienced developers are relatively unaffected by identifier style, so our recommendation is just to use consistent style in all examples. Since most languages have style guides (e.g., PEP 8[5] for Python) and tools to check that code follows these guidelines, our full recommendation is to *use tools to ensure that all code examples adhere to a consistent style.*

---

[5]https://www.python.org/dev/peps/pep-0008/

7.9 DO BETTER ERROR MESSAGES HELP?

Incomprehensible error messages are a major source of frustration for novices (and for experienced programmers as well). Several researchers have therefore explored whether better error messages would help alleviate this. For example, [Beck2016] rewrote some of the Java compiler's messages so that instead of:

```
C:\stj\Hello.java:2: error: cannot find symbol
        public static void main(string[ ] args)
^
1 error
Process terminated ... there were problems.
```

learners would see:

```
Looks like a problem on line number 2.
If "string" refers to a datatype, capitalize the 's'!
```

Sure enough, novices given these messages made fewer repeated errors and fewer errors overall.

[Bari2017] went further and used eye tracking to show that despite the grumblings of compiler writers, people really do read error messages—in fact, they spend 13–25% of their time doing this. However, reading error messages turns out to be as difficult as reading source code, and how difficult it is to read the error messages strongly predicts task performance. Teachers should therefore *show learners how to read and interpret error messages*. [Marc2011] has a rubric for responses to error messages that can be useful in grading such exercises.

DOES VISUALIZATION HELP?

Visualizing program execution is a perennially popular idea, and tools like the Online Python Tutor [Guo2013] and Loupe[6] (which shows how JavaScript's event loop works) are useful teaching aids. However, people learn more from constructing visualizations than they do from viewing visualizations constructed by others [Stas1998; Ceti2016], so does visualization actually help learning?

To answer this, [Cunn2017] replicated an earlier study of the kinds of sketching learners do when tracing code execution. They found that not sketching at all correlates with lower success, while tracing changes to variables' values by writing new values near their names as they change was the most effective strategy.

One possible confounding effect they checked was time: since sketchers take significantly more time to solve problems, do they do better just because they think for longer? The answer is no: there was no correlation between the time taken and the score achieved. Our recommendation is therefore to *teach learners to trace variables' values when debugging*.

---

[6]http://latentflip.com/loupe/

Flowcharts

One often-overlooked finding about visualization is that people under-
stand flowcharts better than pseudocode if both are equally well structured *[Scan1989]. Earlier work showing that pseudocode outperformed*
flowcharts used structured pseudocode and tangled flowcharts; when the
playing field was leveled, novices did better with the graphical representa-
tion.

7.10   WHAT ELSE CAN WE DO TO HELP?

[Viha2014] examined the average improvement in pass rates of various kinds of inter-
vention in programming classes. They point out that there are many reasons to take
their findings with a grain of salt: the pre-change teaching practices are rarely stated
clearly, the quality of change is not judged, and only 8.3% of studies reported negative findings, so either there is positive reporting bias or the way we're teaching right
now is the worst possible and anything would be an improvement. And like many
other studies discussed in this chapter, they were only looking at university classes,
so their findings may not generalize to other groups.

With those caveats in mind, they found ten things teachers can do to improve
outcomes (Figure 7.3):

Collaboration: Activities that encourage learner collaboration either in classrooms
or labs.

Content Change: Parts of the teaching material were changed or updated.

Contextualization: Course content and activities were aligned towards a specific
context such as games or media.

CS0: Creation of a preliminary course to be taken before the introductory programming course; could be organized only for some (e.g., at-risk) learners.

Game Theme: A game-themed component was introduced to the course.

Grading Scheme: A change in the grading scheme, such as increasing the weight
of programming activities while reducing that of the exam.

Group Work: Activities with increased group work commitment such as team-
based learning and cooperative learning.

Media Computation: Activities explicitly declaring the use of media computation
(Chapter 10).

Peer Support: Support by peers in form of pairs, groups, hired peer mentors or
tutors.

Other Support: An umbrella term for all support activities, e.g., increased teacher
hours, additional support channels, etc.

This list highlights the importance of cooperative learning. [Beck2013] looked at
this specifically over three academic years in courses taught by two different teachers and found significant benefits overall and for many subgroups. The cooperators

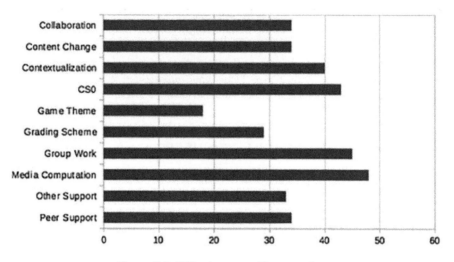

Figure 7.3: Effectiveness of interventions

had higher grades and left fewer questions blank on the final exam, which indicates greater self-efficacy and willingness to try to debug things.

Computing Without Coding

*Writing code isn't the only way to teach people how to program: having novices work on computational creativity exercises improves grades at several levels [Shel2017]. A typical exercise is to describe an everyday object (such as a paper clip or toothbrush) in terms of its inputs, outputs, and functions. This kind of teaching is sometimes called **unplugged**; the CS Unplugged[7] site has lessons and exercises for doing this.*

7.11   WHERE NEXT?

For those who want to go deeper, [Finc2019] is a comprehensive summary of CER, [Ihan2016] summarizes the methods that studies use most often. I hope that some day we will have catalogs like [Ojos2015] and more teacher training materials like [Hazz2014; Guzd2015a; Sent2018] to help us all do better.

Most of the research reported in this chapter was publicly funded but is locked away behind paywalls: at a guess, I broke the law 250 times to download papers from sites like Sci-Hub[8] while writing this book. I hope the day is coming when no one will need to do that; if you are a researcher, please hasten that day by publishing your work in open access venues.

---

[7]https://csunplugged.org/en/

[8]https://en.wikipedia.org/wiki/Sci-Hub

7.12 EXERCISES

YOUR LEARNERS' MISUNDERSTANDINGS (SMALL GROUPS/15)

Working in small groups, re-read Section 7.3 and make a list of misconceptions you think your learners have. How specific are they? How would you check how accurate your list is?

CHECKING FOR COMMON ERRORS (INDIVIDUAL/20)

These common errors are taken from a longer list in [Sirk2012]:

Inverted assignment: The learner assigns the value of the left-hand variable to the right-hand variable rather than the other way around.

Wrong branch: The learner thinks the code in the body of an if is run even if the condition is false.

Executing function instead of defining it: The learner believes that a function is executed as it is defined.

Write one exercise for each to check that learners *aren't* making that mistake.

MANGLED CODE (PAIRS/15)

[Chen2017] describes exercises in which learners reconstruct code that has been mangled by removing comments, deleting or replacing lines of code, moving lines, and so on. Performance on these correlates strongly with performance on assessments in which learners write code, but these questions require less work to mark. Take the solution to a programming exercise you've created in the past, mangle it in two different ways, swap with a partner, and see how long it takes each of you to answer the other's question correctly.

THE RAINFALL PROBLEM (PAIRS/10)

[Solo1986] introduced the Rainfall Problem, which has been used in many subsequent studies of programming [Fisl2014; Simo2013; Sepp2015]. Write a program that repeatedly reads in positive integers until it reads the integer 99999. After seeing 99999, the program prints the average of the numbers seen.

1. Solve the Rainfall Problem in the programming language of your choice.
2. Compare your solution with that of your partner. What does yours do that theirs doesn't and vice versa?

ROLES OF VARIABLES (PAIRS/15)

[Kuit2004; Byck2005; Saja2006] presented a set of single-variable patterns that I have found very useful in teaching beginners:

Fixed value: A data item that does not get a new proper value after its initialization.

Stepper: A data item stepping through a systematic, predictable succession of values.

Walker: A data item traversing in a data structure.

Most-recent holder: A data item holding the latest value encountered while going through a succession of values.

Most-wanted holder: A data item holding the best or most appropriate value encountered so far.

Gatherer: A data item accumulating the effect of individual values.

Follower: A data item that always gets its new value from the old value of some other data item.

One-way flag: A two-valued data item that cannot get its initial value once the value has been changed.

Temporary: A data item holding some value for a very short time only.

Organizer: A data structure storing elements that can be rearranged.

Container: A data structure storing elements that can be added and removed.

Choose a 5–15 lines program and classify its variables using these categories. Compare your classifications with those of a partner. Where you disagreed, did you understand each other's view?

WHAT ARE YOU TEACHING? (INDIVIDUAL/10)

Compare the topics you teach to the list developed in [Luxt2017] (Section 7.1). Which topics do you cover? Which *don't* you cover? What extra topics do you cover that aren't in their list?

BENEFICIAL ACTIVITIES (INDIVIDUAL/10)

Look at the list of interventions developed by [Viha2014] (Section 7.10). Which of these things do you already do in your classes? Which ones could you easily add? Which ones are irrelevant?

MISCONCEPTIONS AND CHALLENGES (SMALL GROUPS/15)

The Professional Development for CS Principles Teaching[9] site includes a detailed list of learners' misconceptions and exercises[10]. Working in small groups, choose one section (such as data structures or functions) and go through their list. Which of these misconceptions do you remember having when you were a learner? Which do you still have? Which have you seen in your learners?

---

[9]http://www.pd4cs.org/

[10]http://www.pd4cs.org/mc-index/

WHAT DO YOU CARE MOST ABOUT? (WHOLE CLASS/15)

[Denn2019] asked people to propose and rate various CER questions, and found that there was no overlap between those that researchers cared most about and those that non-researchers cared most about. The researchers' favorites were:

1. What fundamental programming concepts are the most challenging for students?
2. What teaching strategies are most effective when dealing with a wide range of prior experience in introductory programming classes?
3. What affects students' ability to generalize from simple programming examples?
4. What teaching practices are most effective for teaching computing to children?
5. What kinds of problems do students in programming classes find most engaging?
6. What are the most effective ways to teach programming to various groups?
7. What are the most effective ways to scale computing education to reach the general student population?

while the most important questions for non-researchers were:

1. How and when is it best to give students feedback on their code to improve learning?
2. What kinds of programming exercises are most effective when teaching students Computer Science?
3. What are the relative merits of project-based learning, lecturing, and active learning for students learning computing?
4. What is the most effective way to provide feedback to students in programming classes?
5. What do people find most difficult when breaking problems down into smaller tasks while programming?
6. What are the key concepts that students need to understand in introductory computing classes?
7. What are the most effective ways to develop computing competency among students in non-computing disciplines?
8. What is the best order in which to teach basic computing concepts and skills?

Have each person in the class independently give one point to each of the eight questions from the combined lists that they care most about, then calculate an average score for each question. Which ones are most popular in your class? In which group are the most popular questions?

8 Teaching as a Performance Art

In *Darwin Among the Machines*, George Dyson wrote, "In the game of life and evolution there are three players at the table: human beings, nature, and machines. I am firmly on the side of nature. But nature, I suspect, is on the side of the machines…" There are similarly now three players in the game of education: textbooks and other read-only materials, live lectures, and automated online lessons. You may give your learners both written lessons and some combination of recorded video and self-paced exercises, but if you are going to teach in person you must offer something different from (and hopefully better than) either of them. This chapter therefore focuses on how to teach programming by actually doing it.

8.1 LIVE CODING

Teaching is theater, not cinema.
— Neal Davis

The most effective way to teach programming is **live coding** [Rubi2013; Haar2017]. Instead of presenting pre-written material, the teacher writes code in front of the class while the learners follow along, typing it in and running it as they go. Live coding works better than slides for several reasons:

- It enables **active teaching** by allowing teachers to follow their learners' interests and questions in the moment. A slide deck is like a railway track: it may be a smooth ride, but you have to decide in advance where you're going. Live coding is more like driving an off-road vehicle: it may be bumpier, but it's a lot easier to change direction and go where people want.
- Watching a program being written is more motivating than watching someone page through slides.
- It facilitates unintended knowledge transfer: people learn more than we are consciously teaching by watching *how* we do things.
- It slows the teacher down: if they have to type in the program as they go along then they can only go twice as fast as their learners rather than ten times faster as they could with slides.
- It helps reduce the load on short-term memory because it makes the teacher more aware of how much they are throwing at their learners.
- Learners get to see how to diagnose and correct mistakes. They are going to spend a lot of time doing this; unless you're a perfect typist, live coding ensures that they get to see how to.

- Watching teachers make mistakes shows learners that it's all right to make mistakes of their own. If the teacher isn't embarrassed about making and talking about mistakes, learners will be more comfortable doing so too.

Another benefit of live coding is that it demonstrates the order in which programs should be written. When looking at how people solved Parsons Problems, [Ihan2011] found that experienced programmers often dragged the method signature to the beginning, then added the majority of the control flow (i.e., loops and conditionals), and only then added details like variable initialization and handling of corner cases. This out-of-order authoring is foreign to novices, who read and write code in the order it's presented on the page; seeing it helps them learn to decompose problems into subgoals that can be tackled one by one. Live coding also gives teachers a chance to emphasize the importance of small steps with frequent feedback [Blik2014] and the importance of picking a plan rather than making more-or-less random changes and hoping things will get better [Spoh1985].

It takes a bit of practice to get comfortable talking while you code in front of an audience, but most people report that it quickly becomes no more difficult than talking around a deck of slides. The sections below offer tips on how to make your live coding better.

EMBRACE YOUR MISTAKES

The typos are the pedagogy.
— Emily Jane McTavish

The most important rule of live coding is to embrace your mistakes. No matter how well you prepare, you will make some; when you do, think through them with your audience. While data is hard to come by, professional programmers spend anywhere from 25% to 60% of their time debugging; novices spend much more (Section 7.6), but most textbooks and tutorials spend little time diagnosing and correcting problems. If you talk aloud while you figure out what you mistyped or where you took the wrong path, and explain how you've corrected yourself, you will give your learners a toolbox they can use when they make their own mistakes.

Deliberate Fumbles

*Once you have given a lesson several times, you're unlikely to make anything other than basic typing mistakes (which can still be informative). You can try to remember past mistakes and make them deliberately, but that often feels forced. An alternative approach is **twitch coding**: ask learners one by one to tell you what to type next. This is pretty much guaranteed to get you into some kind of trouble.*

ASK FOR PREDICTIONS

One way to keep learners engaged while you are live coding is to ask them to make predictions about what the code on the screen is going to do. You can then write down the first few suggestions they make, have the whole class vote on which they think is most likely, and then run the code. This is a lightweight form of peer instruction, which we will discuss in Section 9.2; as well as keeping their attention on task, it gives them practice at reasoning about code's behavior.

TAKE IT SLOW

Every time you type a command, add a line of code to a program, or select an item from a menu, say what you are doing out loud and then point to what you have done and its output on the screen and go through it a second time. This allows learners to catch up and to check that what they have just done is correct. It is particularly important when some of your learners have seeing or hearing challenges or are not fluent in the language of instruction.

Whatever you do, *don't* copy and paste code: doing this practically guarantees that you'll race ahead of your learners. And if you use tab completion, say it out loud so that your learners understand what you're doing: "Let's use turtle dot 'r' 'i' and tab to get 'right'."

If the output of your command or code makes what you just typed disappear from view, scroll back up so learners can see it again. If that's not practical, execute the same command a second time or copy and paste the last command(s) into the workshop's shared notes.

BE SEEN AND HEARD

When you sit down, you are more likely to look at your screen rather than at your audience and may be hidden from learners in the back rows of your classroom. If you are physically able to stand up for a couple of hours, you should therefore do so while teaching. Plan for this and make sure that you have a raised table, standing desk, or lectern for your laptop so that you don't have to bend over to type.

Regardless of whether you are standing or sitting, make sure to move around as much as you can: go to the screen to point something out, draw something on the whiteboard, or just step away from your computer for a few moments and speak directly to your audience. Doing this draws your learners' attention away from their screens and provides a natural time for them to ask questions.

If you are going to be teaching for more than a couple of hours, it's worth using a microphone even in a small room. Your throat gets tired just like every other part of your body; using a mike is no different from wearing comfortable shoes (which you also ought to do). It can also make a big difference to people with hearing disabilities.

MIRROR YOUR LEARNER'S ENVIRONMENT

You may have customized your environment with a fancy Unix shell prompt, a custom color scheme for your development environment, or a plethora of keyboard shortcuts. Your learners won't have any of this, so try to create an environment that mirrors what they *do* have. Some teachers create a separate bare-bones account on their laptop or a separate teaching-only account if they're using an online service like Scratch or GitHub. Doing this can also help prevent packages that you installed for work yesterday breaking the lesson you are supposed to teach this morning.

USE THE SCREEN WISELY

You will usually need to enlarge your font considerably for people to read it from the back of the room, which means you can put much less on the screen than you're used to. In many cases you will be reduced to 60–70 columns and 20–30 rows, so that you're using a 21st century supercomputer as if it was a dumb terminal from the early 1980s.

To manage this, maximize the window you are using to teach and then ask everyone to give you a thumbs-up or thumbs-down on its readability. Use a black font on a lightly-tinted background rather than a light font on a dark background—the light tint will glare less than pure white.

Pay attention to the room lighting as well: it should not be fully dark, and there should be no lights directly on or above your projection screen. Allow a few minutes for learners to reposition their tables so that they can see clearly.

When the bottom of the projection screen is at the same height as your learners' heads, people in the back won't be able to see the lower parts. You can raise the bottom of your window to compensate, but this will give you even less space for typing.

If you can get a second projector and screen, use it: the extra real estate will allow you to display your code on one side and its output or behavior on the other. If second screen requires its own computer, ask a helper to control it rather than hopping back and forth between two keyboards.

Finally, if you are teaching something like the Unix shell in a console window, it's important to tell people when you run an in-console text editor and when you return to the console prompt. Most novices have never seen a window take on multiple personalities in this way, and can quickly become confused by when you are interacting with the shell, when you are typing in the editor, and when you are dealing with an interactive prompt for Python or some other language. You can avoid this problem by using a separate window for editing; if you do this, always tell learners when you are switching focus from one to the other.

Accessibility Aids Help Everyone
Tools like Mouseposé[1] (for Mac) and PointerFocus[2] (for Windows) highlight the position of your mouse cursor on the screen, and screen recording tools like Camtasia[3] and standalone applications like KeyCastr[4] echo

---

[1] https://boinx.com/mousepose/overview/

[2] http://www.pointerfocus.com/

[3] https://www.techsmith.com/video-editor.html

[4] https://github.com/keycastr/keycastr

invisible keys like tab and Control-J as you type them. These can be a bit annoying when you first start to use them, but help your learners figure out what you're doing.

DOUBLE DEVICES

Some people now use two devices when teaching: a laptop plugged into the projector for learners to see and a tablet so that they can view their own notes and the notes that the learners are taking (Section 9.7). This is more reliable than flipping back and forth between virtual desktops, though a printout of the lesson is still the most reliable backup technology.

DRAW EARLY, DRAW OFTEN

Diagrams are always a good idea. I sometimes have a slide deck full of ones that I have prepared in advance, but building diagrams step by step helps with retention (Section 4.1) and allows you to improvise.

AVOID DISTRACTIONS

Turn off any notifications you may use on your laptop, especially those from social media. Seeing messages flash by on the screen distracts you as well as your learners, and it can be awkward when a message pops up you'd rather not have others see. Again, you might want to create a second account on your computer that doesn't have email or other tools set up at all.

IMPROVISE—AFTER YOU KNOW THE MATERIAL

Stick fairly closely to the lesson plan you've drawn up or borrowed the first time you teach a lesson. It may be tempting to deviate from the material because you would like to show a neat trick or demonstrate another way to do something, but there is a fair chance you'll run into something unexpected that will cost you more time than you have.

Once you are more familiar with the material, though, you can and should start improvising based on the backgrounds of your learners, their questions in class, and what you personally find most interesting. This is like playing a new song: you stick to the sheet music the first few times, but after you're comfortable with the melody and chord changes, you can start to put your own stamp on it.

When you want to use something new, run through it beforehand *using the same computer that you'll be teaching on*: installing several hundred megabytes of software over high school WiFi in front of bored 16-year-olds isn't something you ever want to have to do.

Direct Instruction

Direct Instruction (DI) is a teaching method centered around meticulous curriculum design delivered through a prescribed script. It's more like an

actor reciting lines than it is like the improvisational approach we rec-
ommend. [Stoc2018] found a statistically significant positive effect for
DI even though it sometimes gets knocked for being mechanical. I prefer
improvisation because DI requires more up-front preparation than most
free-range learning groups can afford.

FACE THE SCREEN—OCCASIONALLY

It's OK to face the projection screen occasionally when you are walking through a section of code or drawing a diagram: *not* looking at a roomful of people who are all looking at you can help lower your anxiety levels and give you a moment to think about what to say next.

You shouldn't do this for more than a few seconds at a time, though. A good rule of thumb is to treat the projection screen as one of your learners: if it would be uncomfortable to stare at someone for as long as you are spending looking at the screen, it's time to turn around and face your class again.

DRAWBACKS

Live coding does have some drawbacks, but they can all be avoided or worked around with a little bit of practice. If you find that you are making too many trivial typing mistakes, set aside five minutes every day to practice typing: it will help your day-to-day work as well. If you think you are spending too much time referring to your lesson notes, break them into smaller pieces so that you only ever have to think about one small step at a time.

And if you feel that you're spending too much time typing in library import statements, class headers, and other boilerplate code, give yourself and your learners some skeleton code as a starting point (Section 9.9). Doing this will also reduce their cognitive load, since it will focus their attention where you want it.

8.2   LESSON STUDY

From politicians to researchers and teachers themselves, educational reformers have designed systems to find and promote people who can teach well and eliminate those who cannot. But the assumption that some people are born teachers is wrong: instead, like any other performance art, the keys to better teaching are practice and collaboration. As [Gree2014] explains, the Japanese approach to this is called **jugyokenkyu**, which means "lesson study":

> In order to graduate, [Japanese] education majors not only had to watch their assigned master teacher work, they had to effectively replace him, installing themselves in his classroom first as observers and then, by the third week, as a wobbly... approximation of the teacher himself. It worked like a kind of teaching relay. Each trainee took a subject, planning five days' worth of lessons... [and then] each took a day. To pass the baton, you had to teach a day's lesson in

every single subject: the one you planned and the four you did not... and you had to do it right under your master teacher's nose. Afterward, everyone—the teacher, the college students, and sometimes even another outside observer— would sit around a formal table to talk about what they saw.

Putting work under a microscope in order to improve it is commonplace in fields as diverse as manufacturing[5] and music. A professional musician, for example, will dissect half a dozen recordings of "Body and Soul" or "Smells Like Teen Spirit" before performing it. They would also expect to get feedback from fellow musicians during practice and after performances.

But continuous feedback isn't part of teaching culture in most English-speaking countries. There, what happens in the classroom stays in the classroom: teachers don't watch each other's lessons on a regular basis, so they can't borrow each other's good ideas. Teachers may get lesson plans and assignments from colleagues, the school board or a textbook publisher, or go through a few MOOCs on the Internet, but each one has to figure out how to deliver specific lessons in specific classrooms for specific learners. This is particularly true for volunteers and other free-range teachers involved in after-school workshops and bootcamps.

Writing up new techniques and giving **demonstration lessons** (in which one person teaches actual learners while other teachers observe) are not solutions. For example, [Finc2007; Finc2012] found that of the 99 change stories analyzed, teachers only searched actively for new practices or materials in three cases, and only consulted published material in eight. Most changes occurred locally, without input from outside sources, or involved only personal interaction with other educators. [Bark2015] found something similar:

> Adoption is not a "rational action"... but an iterative series of decisions made in a social context, relying on normative traditions, social cueing, and emotional or intuitive processes... Faculty are not likely to use educational research findings as the basis for adoption decisions... Positive student feedback is taken as strong evidence by faculty that they should continue a practice.

Jugyokenkyu works because it maximizes the opportunity for unplanned knowledge transfer between teachers: someone sets out to demonstrate X, but while watching them, their audience actually learns Y as well (or instead). For example, a teacher might intend to show learners how to search for email addresses in a text file, but what their audience might take away is some new keyboard shortcuts.

8.3 GIVING AND GETTING FEEDBACK ON TEACHING

Observing someone helps you, and giving them feedback helps them, but it can be hard to receive feedback, especially when it's negative (Figure 8.1).

---

[5]https://en.wikipedia.org/wiki/W._Edwards_Deming

Figure 8.1: Feedback feelings (copyright © Deathbulge 2013)

Feedback is easier to give and receive when both parties share expectations about what is and isn't in scope and about how comments ought to be phrased. If you are the person asking for feedback:

Initiate feedback. It's better to ask for feedback than to receive it unwillingly.

Choose your own questions, i.e., ask for specific feedback. It's a lot harder for someone to answer, "What do you think?" than to answer either, "Was I speaking too quickly?" or , "What is one thing from this lesson I should keep doing?" Directing feedback like this is also more helpful to you. It's always better to try to fix one thing at once than to change everything and hope it's for the better. Directing feedback at something you have chosen to work on helps you stay focused, which in turn increases the odds that you'll see progress.

Use a feedback translator. Have someone else read over all the feedback and give you a summary. It can be easier to hear, "Several people think you could speed up a little," than to read several notes all saying, "This is too slow" or, "This is boring."

Be kind to yourself. Many of us are very critical of ourselves, so it's always helpful to jot down what we thought of ourselves *before* getting feedback from others. That allows us to compare what we think of our performance with what others think, which in turn allows us to scale the former more accurately. For example, it's very common for people to think that they're saying "um" and "err" too often when their audience doesn't notice it. Getting that feedback once allows teachers to adjust their assessment of themselves the next time they feel that way.

You can give feedback to others more effectively as well:

Interact. Staring at someone is a good way to make them feel uncomfortable, so if you want to give feedback on how someone normally teaches, you need to set

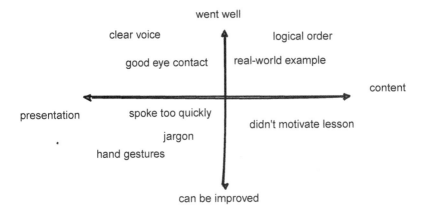

Figure 8.2: Teaching rubric

them at ease. Interacting with them the way that a real learner would is a good way to do this, so ask questions or (pretend to) type along with their example. If you are part of a group, have one or two people play the role of learner while the others take notes.

Balance positive and negative feedback. The "compliment sandwich" made up of one positive comment, one negative, and a second positive becomes tiresome pretty quickly, but it's important to tell people what they should keep doing as well as what they should change[6].

Take notes. You won't remember everything you noticed if the presentation lasts longer than a few seconds, and you definitely won't recall how often you noticed them. Make a note the first time something happens and then add a tick mark when it happens again so that you can sort your feedback by frequency.

Taking notes is more efficient when you have some kind of rubric so that you're not scrambling to write your observations while the person you're observing is still talking. The simplest rubric for free-form comments from a group is a 2x2 grid whose vertical axis is labeled "what went well" and "what can be improved", and whose horizontal axis is labeled "content" (what was said) and "presentation" (how it was said). Observers write their comments on sticky notes as they watch the demonstration, then post those in the quadrants of a grid drawn on a whiteboard (Figure 8.2).

Rubrics and Question Budgets
Section F.1 contains a sample rubric for assessing 5–10 minutes of programming instruction. It presents items in more or less the order that they're likely to come up, e.g., questions about the introduction come before questions about the conclusion.

---

[6]For a while, I was so worried about playing in tune that I completely lost my sense of timing.

> *Rubrics like this one tend to grow over time as people think of things they'd like to add. A good way to keep them manageable is to insist that the total length stays constant: if someone wants to add a question, they have to identify one that's less important and can be removed.*

If you are interested in giving and getting feedback, [Gorm2014] has good advice that you can use to make peer-to-peer feedback a routine part of your teaching, while [Gawa2011] looks at the value of having a coach.

> ***Studio Classes***
> *Architecture schools often include studio classes in which students solve small design problems and get feedback from their peers right then and there. These classes are most effective when the teacher critiques the peer critiques so that participants learn not only how to make buildings but how to give and get feedback [Scho1984]. Master classes in music serve a similar purpose, and I have found that giving feedback on feedback helps people improve their teaching as well.*

8.4   HOW TO PRACTICE PERFORMANCE

The best way to improve your in-person lesson delivery is to watch yourself do it:

• Work in groups of three.
• Each person rotates through the roles of teacher, audience, and videographer. The teacher has 2 minutes to explain something. The person pretending to be the audience is there to be attentive, while the videographer records the session using a cellphone or other handheld device.
• After everyone has finished teaching, the whole group watches the videos together. Everyone gives feedback on all three videos, i.e., people give feedback on themselves as well as on others.
• After the videos have been discussed, they are deleted. (Many people are justifiably uncomfortable about images of themselves appearing online.)
• Finally, the whole class reconvenes and adds all the feedback to a shared 2x2 grid of the kind described above *without* saying who each item of feedback is about.

In order for this exercise to work well:

• Record all three videos and then watch all three. If the cycle is teach-review-teach-review, the last person to teach invariably runs short of time (sometimes on purpose). Doing all the reviewing after all the teaching also helps put a bit of distance between the two, which makes the exercise slightly less excruciating.
• Let people know at the start of the class that they will be asked to teach something so that they have time to choose a topic. Telling them this too far in advance can be counter-productive, since some people will fret over how much they should prepare.

- Groups must be physically separated to reduce audio cross-talk between their recordings. In practice, this means 2–3 groups in a normal-sized classroom, with the rest using nearby breakout spaces, coffee lounges, offices, or (on one occasion) a janitor's storage closet.
- People must give feedback on themselves as well as on each other so that they can calibrate their impressions of their own teaching against those of other people. Most people are harder on themselves than they ought to be, and it's important for them to realize this.

The announcement of this exercise is often greeted with groans and apprehension, since few people enjoy seeing or hearing themselves. However, those same people consistently rate it as one of the most valuable parts of teaching workshops. It's also good preparation for co-teaching (Section 9.3): teachers find it a lot easier to give each other informal feedback if they have had some practice doing so and have a shared rubric to set expectations.

And speaking of rubrics: once the class has put all of their feedback on a shared grid, pick a handful of positive and negative comments, write them up as a checklist, and have them do the exercise again. Most people are more comfortable the second time around, and being assessed on the things that they themselves have decided are important increases their sense of self-determination (Chapter 10).

Tells

We all have nervous habits: we talk more rapidly and in a higher-pitched voice than usual when we're on stage, play with our hair, or crack our knuckles. Gamblers call these "tells," and people often don't realize that they pace, look at their shoes, or rattle the change in their pocket when they don't actually know the answer to a question.

You can't get rid of tells completely, and trying to do so can make you obsess about them. A better strategy is to try to displace them—for example, to train yourself to scrunch your toes inside your shoes when you're nervous instead of cleaning your ear with your pinky finger.

8.5   EXERCISES

GIVE FEEDBACK ON BAD TEACHING (WHOLE CLASS/20)

As a group, watch this video of bad teaching[7] and give feedback on two axes: positive versus negative and content versus presentation. Have each person in the class add one point to a 2x2 grid on a whiteboard or in the shared notes without duplicating any points. What did other people see that you missed? What did they think that you strongly agree or disagree with?

---

[7] https://www.youtube.com/watch?v=-ApVt04rB4U

PRACTICE GIVING FEEDBACK (SMALL GROUPS/45)

Use the process described in Section 8.4 to practice teaching in groups of three and pool feedback.

THE BAD AND THE GOOD (WHOLE CLASS/20)

Watch the videos of live coding done poorly[8] and live coding done well[9] and summarize your feedback on the differences using the usual 2x2 grid. How is the second round of teaching better than the first? Is there anything that was better in the first than in the second?

SEE, THEN DO (PAIRS/30)

Teach 3–4 minutes of a lesson using live coding to a classmate, then swap and watch while that person live codes for you. Don't bother trying to record these sessions—it's difficult to capture both the person and the screen with a handheld device—but give feedback the same way you have previously. Explain in advance to your fellow trainee what you will be teaching and what the learners you teach it to are expected to be familiar with.

- What felt different about live coding compared to standing up and lecturing? What was easier or harder?
- Did you make any mistakes? If so, how did you handle them?
- Did you talk and type at the same time, or alternate?
- How often did you point at the screen? How often did you highlight with the mouse?
- What will you try to keep doing next time? What will you try to do differently?

TELLS (SMALL GROUPS/15)

1. Make a note of what you think your tells are, but do not share them with other people.
2. Teach a short lesson (2–3 minutes long).
3. Ask your audience how they think you betray nervousness. Is their list the same as yours?

TEACHING TIPS (SMALL GROUPS/15)

The CS Teaching Tips[10] site has a large number of practical tips on teaching computing, as well as a collection of downloadable tip sheets. Go through their tip sheets

---

[8] https://youtu.be/bXxBeNkKmJE

[9] https://youtu.be/SkPmwe_WjeY

[10] http://csteachingtips.org/

in small groups and classify each tip according to whether you use it all the time, use it occasionally, or never use it. Where do your practice and your peers' practice differ? Are there any tips you strongly disagree with or think would be ineffective?

REVIEW

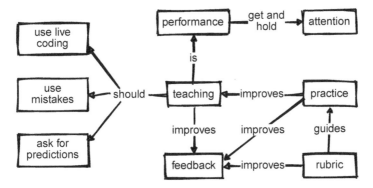

Figure 8.3: Concepts: Feedback

9 In the Classroom

The previous chapter described how to practice lesson delivery and described one method—live coding—that allows teachers to adapt to their learners' pace and interests. This chapter describes other practices that are also useful in programming classes.

Before describing them, it's worth pausing for a moment to set expectations. The best teaching method we know is individual tutoring: [Bloo1984] found that students taught one-to-one did two standard deviations better than those who learned through conventional lecture, i.e., that individually tutored students outperformed 98% of students who were lectured to. However, while mentoring and apprenticeship were the most common ways to pass on knowledge throughout most of history, the industrialization of formal education has made it the exception today. Despite the hype around artificial intelligence, it isn't going to square this circle any time soon, so every method described below is essentially an attempt to approach the effectiveness of individual tutoring at scale.

9.1 ENFORCE THE CODE OF CONDUCT

The most important thing I've learned about teaching in the last 30 years is how important it is for everyone to treat everyone else with respect, both in and out of class. If you use this material in any way, please adopt a Code of Conduct like the one in Appendix B and require everyone who takes part in your classes to abide by it. It can't stop people from being offensive, any more than laws against theft stop people from stealing, but it *can* make expectations and consequences clear, and signal that you are trying to make your class welcoming to all.

But a Code of Conduct is only useful if it is enforced. If you believe that someone has violated yours, you may warn them, ask them to apologize, and/or expel them, depending on the severity of the violation and whether or not you believe it was intentional. Whatever you do:

Do it in front of witnesses. Most people will tone down their language and hostility in front of an audience, and having someone else present ensures that later discussion doesn't degenerate into conflicting claims about who said what.

If you expel someone, say so to the rest of the class and explain why. This helps prevent rumors from spreading and shows that your Code of Conduct actually means something.

Email the offender as soon as you can to summarize what happened and the steps you took, and copy the message to your workshop's hosts or one of your fellow teachers so that there's a contemporaneous record of the conversation. If the offender replies, don't engage in a long debate: it's never productive.

What happens outside of class matters at least as much as what happens within it [Part2011], so you need to provide a way for learners to report problems that you aren't there to see yourself. One step is to ask someone who isn't part of your group to be the first point of contact; that way, if someone wants to make a complaint about you or one of your fellow teachers, they have some assurance of confidentiality and independent action. [Auro2019] has lots of other advice and is both short and practical.

9.2 PEER INSTRUCTION

No matter how good a teacher is, they can only say one thing at a time. How then can they clear up many different misconceptions in a reasonable time? The best solution developed so far is a technique called **peer instruction**. Originally created by Eric Mazur at Harvard [Mazu1996], it has been studied extensively in a wide variety of contexts, including programming [Crou2001; Port2013], and [Port2016] found that learners value peer instruction even at first contact.

Peer instruction attempts to provide one-to-one instruction in a scalable way by interleaving formative assessment with learner discussion:

1. Give a brief introduction to the topic.
2. Give learners a multiple choice question that probes for their misconceptions (rather than testing simple factual knowledge).
3. Have all the learners vote on their answers to the MCQ.
    - If the learners all have the right answer, move on.
    - If they all have the same wrong answer, address that specific misconception.
    - If they have a mix of right and wrong answers, give them several minutes to argue with each other in groups of 2–4, then vote again.

As this video[1] shows, group discussion significantly improves learners' understanding because it uncovers gaps in their reasoning and forces them to clarify their thinking. Re-polling the class then lets the teacher know if they can move on or if further explanation is necessary. A final round of additional explanation after the correct answer is presented gives learners one more chance to solidify their understanding.

But could this be a false positive? Are results improving because of increased understanding during discussion or simply from a follow-the-leader effect ("vote like Jane, she's always right")? [Smit2009] tested this by following the first question with a second one that learners answered individually. They found that peer discussion actually does enhance understanding, even when none of the learners in a discussion group originally knew the correct answer. As long as there is a diversity of opinion within the group, their misconceptions cancel out.

---

[1]https://www.youtube.com/watch?v=2LbuoxAy56o

Taking a Stand

It is important to have learners vote publicly so that they can't change their minds afterward and rationalize it by making excuses to themselves like "I just misread the question." Much of the value of peer instruction comes from the hypercorrection of getting the wrong answer and having to think through the reasons why (Section 5.1).

9.3   TEACH TOGETHER

Co-teaching describes any situation in which two teachers work together in the same classroom. [Frie2016] describes several ways to do this:

Team teaching: Both teachers deliver a single stream of content in tandem, taking turns like musicians taking solos.

Teach and assist: Teacher A teaches while Teacher B moves around the classroom to help struggling learners.

Alternative teaching: Teacher A provides a small set of learners with more intensive or specialized instruction while Teacher B delivers a general lesson to the main group.

Teach and observe: Teacher A teaches while Teacher B observes the learners, collecting data on their understanding to help plan future lessons.

Parallel teaching: The class is divided in two and the teachers present the same material simultaneously to each.

Station teaching: The learners are divided into small groups that rotate from one station or activity to the next while teachers supervise where needed.

All of these models create more opportunities for unintended knowledge transfer than teaching alone. Team teaching is particularly beneficial in day-long workshops: it gives each teacher's voice a chance to rest and reduces the risk that they will be so tired by the end of the day that they will start snapping at their learners or fumbling at their keyboard.

Helping

Many people who aren't comfortable teaching are willing and able to provide in-class technical support. They can help learners with setup and installation, answer technical questions during exercises, monitor the room to spot people who may need help, or keep an eye on the shared notes (Section 9.7), and either answer questions or remind the teacher to do so during breaks.

Helpers are sometimes people training to become teachers (i.e., they're Teacher B in the teach and assist model), but they can also be members of the host institution's technical support staff, alumni of the class, or advanced learners who already know the material well. Using the latter as helpers is doubly effective: not only are they more likely to understand the problems their peers are having, it also stops them from getting bored. This helps the whole class stay engaged because boredom is infectious: if a handful of people start checking out, the people around them will follow suit.

If you and a partner are co-teaching:

- Take 2–3 minutes before the start of each class to confirm who's teaching what. If you have time, try drawing or reviewing a concept map together.
- Use that time to work out a couple of hand signals as well. "You're going too fast," "speak up," "that learner needs help," and, "It's time for a bathroom break" are all useful.
- Each person should teach for at least 10–15 minutes at a stretch, since learners will be distracted by more frequent switch-ups.
- The person who isn't teaching shouldn't interrupt, offer corrections or elaborations, or do anything else to distract from what the person teaching is doing or saying. The one exception is to ask leading questions if the learners seem lethargic or unsure of themselves.
- Each person should take a couple of minutes before they start teaching to see what their partner is going to teach after they're done, and then *not* present any of that material.
- The person who isn't teaching should stay engaged with the class, not catch up on their email. Monitoring the shared notes (Section 9.7), keeping an eye on the learners to see who's struggling, jotting down some feedback to give your teaching partner at the next break—anything that contributes to the lesson is better than anything that doesn't.

Most importantly, take a few minutes when the class is over to congratulate or commiserate with each other: in teaching as in life, shared misery is lessened and shared joy increased.

9.4 ASSESS PRIOR KNOWLEDGE

The more you know about your learners before you start teaching, the more you will be able to help them. Inside a formal school system, the prerequisites to your course will tell you something about what they're likely to already know. In a free-range setting, though, your learners may be much more diverse, so you may want to give them a short survey or questionnaire in advance of your class to find out what knowledge and skills they have.

Asking people to rate themselves on a scale from 1 to 5 is pointless because the less people know about a subject, the less accurately they can estimate their knowledge (Figure 9.1, from https://theness.com/neurologicablog/index.php/misunderstanding-dunning-kruger/[2]), a phenomenon called the **Dunning-Kruger**

---

[2]Neurologica

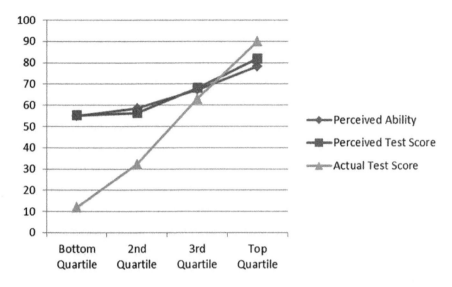

Figure 9.1: The Dunning-Kruger Effect

effect [Krug1999]. Conversely, people who are members of underrepresented groups will often underrate their skills.

Rather than asking people to self-assess, you can ask them how easily they could complete some specific tasks. Doing this is risky, though, because school trains people to treat anything that looks like an exam as something they have to pass rather than as a chance to shape instruction. If someone answers "I don't know" to even a couple of questions on your pre-assessment, they might conclude that your class is too advanced for them. You could therefore scare off many of the people you most want to help.

Section F.5 presents a short pre-assessment questionnaire that most potential learners are unlikely to find intimidating. If you use it or anything like it, try to follow up with people who *don't* respond to find out why not and compare your evaluation of learners with their self-assessment to improve your questions.

9.5 PLAN FOR MIXED ABILITIES

If your learners have widely varying levels of prior knowledge, you can easily wind up in a situation where a third of your class is lost and a third is bored. That's unsatisfying for everyone, but there are some strategies you can use to manage the situation:

- Before running a workshop, communicate its level clearly to everyone by showing a few examples of exercises that they will be asked to complete. This helps potential participants gauge the level of the class far more effectively than a point-form list of topics.

- Provide extra self-paced exercises so that more advanced learners don't finish early and get bored.
- Keep an eye out for learners who are falling behind and intervene early so that they don't become frustrated and give up.
- Ask more advanced learners to help people next to them (see Section 9.6 below).

One other way to accommodate mixed abilities is to have everyone work through material on their own at their own pace as they would in an online course, but to do it simultaneously and side by side with helpers roaming the room to get people unstuck. Some people will go three or four times further than others when workshops are run like this, but everyone will have had a rewarding and challenging day.

False Beginners

*A **false beginner** is someone who has studied a language before but is learning it again. They may be indistinguishable from **absolute beginners** on pre-assessment tests, but they are able to move much more quickly once the class starts because they are re-learning rather than learning for the first time.*

*Being a false beginner is often a sign of **preparatory privilege** [Marg2010], and false beginners are common in free-range programming classes. For example, a child whose family is affluent enough to have sent them to a robotics summer camp may do poorly on a pre-test of programming knowledge because the material isn't fresh in their mind, but still has an advantage over a child from a less fortunate background. The strategies described above can help level the playing field in cases like this, but again, the real solution is to use your own privilege to address larger out-of-class factors [Part2011].*

The most important thing is to accept that you can't help everyone all of the time. If you slow down to accommodate two people who are struggling, you are failing the other 18. Equally, if you spend a few minutes talking about an advanced topic to a learner who is bored, the rest of the class will feel left out.

9.6 PAIR PROGRAMMING

Pair programming is a software development practice in which two programmers work together on one computer[3]. One person (the driver) does the typing while the other (the navigator) offers comments and suggestions, and the two switch roles several times per hour.

Pair programming is an effective practice in professional work [Hann2009] and is also a good way to teach: benefits include increased success rate in introductory courses, better software, and higher learner confidence in their solutions. There is also evidence that learners from underrepresented groups benefit even more than

---

[3]https://www.youtube.com/watch?v=vgkahOzFH2Q

others [McDo2006; Hank2011; Port2013; Cele2018]. Partners can help each other out during practical exercises, clarify each other's misconceptions when the solution is presented, and discuss common interests during breaks. I have found it particularly helpful with mixed-ability classes, since pairs are more homogeneous than individuals.

When you use pairing, put *everyone* in pairs, not just learners who are struggling, so that no one feels singled out. It's also useful to have people sit in new places (and hence pair with different partners) on a regular basis, and to have people switch roles within each pair three or four times per hour so that the stronger personality in each pair doesn't dominate the session.

If your learners are new to pair programming, take a few minutes to demonstrate what it actually looks like so that they understand that the person who doesn't have their hands on the keyboard isn't supposed to just sit and watch. Finally, tell them that people who focus on trying to complete the task as quickly as possible are less fair in their sharing [Lewi2015].

Switching Partners

Teachers have mixed opinions on whether people should be required to change partners at regular intervals. On the one hand it gives everyone a chance to gain new insights and make new friends. On the other, moving computers and power adapters to new desks several times a day is disruptive, and pairing can be uncomfortable for introverts. That said, [Hann2010] found weak correlation between the "Big Five" personality traits and performance in pair programming, although an earlier study [Wall2009] found that pairs whose members had differing levels of personality traits communicated more often.

9.7 TAKE NOTES... TOGETHER?

Note-taking is a form of real-time elaboration (Section 5.1): it forces you to organize and reflect on material as it's coming in, which in turn increases the likelihood that you will transfer it to long-term memory. Many studies have shown that taking notes while learning improves retention [Aike1975; Boha2011]. While it has not yet been widely studied [Ornd2015; Yang2015], I have found that having learners take notes together in a shared online page is also effective:

- It allows people to compare what they think they're hearing with what other people are hearing, which helps them fill in gaps and correct misconceptions right away.
- It gives the more advanced learners in the class something useful to do. Rather than getting bored and checking Instagram during class, they can take the lead in recording what's being said, which keeps them engaged and allows less advanced learners to focus more of their attention on new material.
- The notes the learners take are usually more helpful *to them* than those the teacher would prepare in advance, since the learners are more likely to write down what they actually found new rather than what the teacher predicted would be new.

- Glancing at recent notes while learners are working on an exercise helps the teacher discover that the class missed or misunderstood something.

Is the Pen Mightier than the Keyboard?

[Muel2014] reported that taking notes on a computer is generally less effective than taking notes using pen and paper. While their result was widely shared, [More2019] was unable to replicate it.

If learners are taking notes together, you can also have them paste in short snippets of code and point-form or sentence-length answers to formative assessment questions. To prevent everyone from trying to edit the same couple of lines at the same time, make a list of everyone's name and paste it into the document whenever you want each person to answer a question.

Learners often find that taking notes together is distracting the first time they try it because they have to split their attention between what the teacher is saying and what their peers are writing (Section 4.1). If you are only working with a particular group once, you should therefore heed the advice in Section 9.12 and have them take notes individually.

Points for Improvement

One way to demonstrate to learners that they are learning with you, not just from you, is to allow them to take notes by editing (a copy of) your lesson. Instead of posting PDFs for them to download, create editable copies of your slides, notes, and exercises in a wiki, a Google Doc, or anything else that allows you to review and comment on changes. Giving people credit for fixing mistakes, clarifying explanations, adding new examples, and writing new exercises doesn't reduce your workload, but increases engagement and the lesson's lifetime (Section 6.3).

9.8   STICKY NOTES

Sticky notes are one of my favorite teaching tools, and I'm not alone in loving their versatility, portability, stickability, foldability, and subtle yet alluring aroma [Ward2015].

AS STATUS FLAGS

Give each learner two sticky notes of different colors, such as orange and green. These can be held up for voting, but their real use is as status flags. If someone has completed an exercise and wants it checked, they put the green sticky note on their laptop; if they run into a problem and need help, they put up the orange one. This works much better than having people raise their hands: it's more discreet (which means they're more likely to actually do it), they can keep working while their flag is raised rather than trying to type one-handed, and the teacher can quickly see from the front of the room what state the class is in. Status flags are particularly helpful when people in mixed-ability classes are working through material at their own speed (Section 9.5).

Once your learners are comfortable with two stickies, give them a third that they can put up when their brains are full or they need a bathroom break[4]. Again, adults are more likely to post a sticky than to raise their hand, and once one blue sticky note goes up, a flurry of others usually follows.

TO DISTRIBUTE ATTENTION

Sticky notes can also be used to ensure the teacher's attention is fairly distributed. Have each learner write their name on a sticky note and put it on their laptop. Each time the teacher calls on them or answers one of their questions, they take their sticky note down. Once all the sticky notes are down, everyone puts theirs up again.

This technique makes it easy for the teacher to see who they haven't spoken with recently, which in turn helps them avoid unconscious bias and interacting preferentially with their most extroverted learners. Without a check like this, it's all too easy to create a feedback loop in which extroverts get more attention, which leads to them doing better, which in turn leads to them getting more attention, while quieter, less confident, or marginalized learners are left behind [Alvi1999; Juss2005].

It also shows learners that attention is being distributed fairly so that when they *are* called on, they won't feel like they're being picked on. When I am working with a new group, I allow people to take down their own sticky notes during the first hour or two of class if they would rather not be called on. If they keep doing this as time goes on, I try to have a quiet conversation with them to find out why and to see if there's anything I can do to make them more comfortable.

AS MINUTE CARDS

You can also use sticky notes as **minute cards**. Before each break, learners take a minute to write one thing on the green sticky note that they think will be useful and one thing on the orange note that they found too fast, too slow, confusing, or irrelevant. While they are enjoying their coffee or lunch, review their notes and look for patterns. It takes less than five minutes to see what learners in a 40-person class are enjoying, what they are confused by, what problems they're having, and what questions you have not yet answered.

Learners should not sign their minute cards: they are meant as anonymous feedback. The one-up/one-down technique described in Section 9.11 is a chance for collective, attributable feedback.

---

[4]A colleague once told me that the basic unit of teaching is the bladder. When I said I'd never thought of that, she said, "You've obviously never been pregnant."

9.9 NEVER A BLANK PAGE

Programming workshops and other kinds of classes can be built around a set of independent exercises, developed into a single extended example in stages, or used as a mixed strategy. The two main advantages of independent exercises are that people who fall behind can easily re-synchronize and that lesson developers can add, remove, and rearrange material at will (Section 6.3). A single extended example, on the other hand, will show learners how the bits and pieces they're learning fit together: in educational parlance, it provides more opportunity for them to integrate their knowledge.

Whichever approach you take, novices should never start doing exercises with a blank page or screen, since they often find this intimidating or bewildering. If they have been following along as you do live coding, ask them to add a few more lines or to modify the example you have built up. Alternatively, if they are taking notes together, paste a few lines of starter code into the document for them to extend or modify.

Modifying existing code instead of writing new code from scratch doesn't just give learners structure: it is also closer to what they will do in real life. Keep in mind, though, that learners may be distracted by trying to understand all of the starter code rather than doing their own work. Java's `public static void main()` or a handful of `import` statements at the top of a Python program may make sense to you, but is extraneous load to them (Chapter 4).

9.10 SETTING UP YOUR LEARNERS

Free-range learners often want to bring their own computers and to leave the class with those machines set up to do real work. Free-range teachers should therefore prepare to teach on both Windows and MacOS[5], even though it would be simpler to require learners to use just one.

Common Denominators
If your participants are using different operating systems, try to avoid using features which are specific to just one and point out any that you do use. For example, the "minimize window" controls and behavior on Windows are different from those on MacOS.

No matter how many platforms you have to deal with, put detailed setup instructions on your course website and email learners a couple of days before the workshop starts to remind them to do the setup. A few people will still show up without the required software because they ran into problems, couldn't find time to complete all the steps, or are simply the sort of person who never follows instructions in advance. To detect this, have everyone run some simple command as soon as they arrive and show the teachers the result, then get helpers and other learners to assist people who have run into trouble.

---

[5]"And Linux!" someone shouts from the back of the room.

Virtual Machines

Some people use tools like Docker[6] to put virtual machines on learners' computers so that everyone is working with exactly the same tools, but this introduces a new set of problems. Older or smaller machines simply aren't fast enough to run them, learners struggle to switch back and forth between two different sets of keyboard shortcuts for things like copying and pasting, and even competent practitioners will become confused about what exactly is happening where.

Setting up is so complicated that many teachers prefer to have learners use browser-based tools instead. However, this makes the class dependent on institutional WiFi (which can be of highly variable quality) and doesn't satisfy learners' desire to leave with their own machines ready for real-world use. As cloud-based tools like Glitch[7] and RStudio Cloud[8] become more robust, though, the latter consideration is becoming less important.

One last way to tackle setup issues is to split the class over several days, and to have people install what's required for each day before leaving class on the day before. Dividing the work into chunks makes each one less intimidating, learners are more likely to actually do it, and it ensures that you can start on time for every lesson except the first.

9.11 OTHER TEACHING PRACTICES

None of the smaller practices described below are essential, but all will improve lesson delivery. As with chess and marriage, success in teaching is often a matter of slow, steady progress.

START WITH INTRODUCTIONS

Begin your class by introducing yourself. If you're an expert, tell them a bit about how you got to where you are; if you're only two steps ahead of them, emphasize what you and they have in common. Whatever you say, your goals are to make yourself more approachable and to encourage their belief that they can succeed.

Learners should also introduce themselves to each other. In a class of a dozen, they can do this verbally; in a larger class or if they are strangers to one another, I find it's better to have them each write a line or two about themselves in the shared notes (Section 9.7).

SET UP YOUR OWN ENVIRONMENT

Setting up your environment is just as important as setting up your learners', but more involved. As well as having network access and all the software you're going

---

[6]http://docker.com

[7]https://glitch.com/

[8]http://rstudio.cloud

to use, you should also have a glass of water or a cup of tea or coffee. This helps keep your throat lubricated, but its real purpose is to give you an excuse to pause and think for a couple of seconds when someone asks a hard question or when you lose track of what you were going to say next. You will probably also want some whiteboard pens and a few of the other things described in Section F.3.

One way to keep your day-to-day work from getting in the way of your teaching is to create a separate account on your computer for the latter. Use system defaults for everything in this second account, along with a larger font and a blank screen background, and turn off notifications so that your teaching isn't interrupted by pop-ups.

AVOID HOMEWORK IN ALL-DAY FORMATS

Learners who have spent an entire day programming will be tired. If you give them homework to do after hours, they'll start the next day tired as well, so don't.

DON'T TOUCH THE LEARNER'S KEYBOARD

It's often tempting to fix things for learners, but even if you narrate every step, it's likely to demotivate them by emphasizing the gap between their knowledge and yours. Instead, keep your hands off the keyboard and talk your learners through whatever they need to do: it will take longer, but it's more likely to stick.

REPEAT THE QUESTION

Whenever someone asks a question in class, repeat it back to them before answering to check that you've understood it and to give people who might not have heard it a chance to do so. This is particularly important when presentations are being recorded or broadcast, since your microphone will usually not pick up what other people are saying. Repeating questions back also gives you a chance to redirect the question to something you're more comfortable answering. . .

ONE UP, ONE DOWN

An adjunct to minute cards is to ask for summary feedback at the end of each day. Learners alternately give either one positive or one negative point about the day without repeating anything that has already been said. The ban on repeats forces people to say things they otherwise might not: once all the "safe" feedback has been given, participants will start saying what they really think.

Different Modes, Different Answers

Minute cards (Section 9.8) are anonymous; the alternating up-and-down feedback is not. You should use the two together because anonymity allows both honesty and trolling.

HAVE LEARNERS MAKE PREDICTIONS

Research has shown that people learn more from demonstrations if they are asked to predict what's going to happen [Mill2013]. Doing this fits naturally into live coding: after adding or changing a few lines of a program, ask the class what is going to happen when it runs. If the example is even moderately complex, prediction can serve as a motivating question for a round of peer instruction.

SETTING UP TABLES

You may not have any control over the layout of the desks or tables in the room in which you teach, but if you do, we find it's best to have flat (dinner-style) seating rather than banked (theater-style) seating so that you can reach learners who need help more easily and so that it's easier for learners to pair with one another (Section 9.5). In-floor power outlets so that you don't have to run power cords across the floor make life easier as well as safer, but are still uncommon.

Whatever layout you have, try to make sure that every seat has an unobstructed view of the screen. Good back support is important too, since people are going to be in them for an extended period. Like in-floor power outlets, good classroom seating is still unfortunately uncommon.

COUGH DROPS

If you talk all day to a room full of people, your throat gets raw because you are irritating the epithelial cells in your larynx and pharynx. This doesn't just make you hoarse—it also makes you more vulnerable to infection (which is part of the reason people often come down with colds after teaching).

The best way to protect yourself against this is to keep your throat lined, and the best way to do that is to use cough drops early and often. Good ones will also mask the onset of coffee breath, for which your learners will probably be grateful.

THINK-PAIR-SHARE

Think-pair-share is a lightweight technique that helps people improve ideas through discussion with their peers. Each person starts by thinking individually about a question or problem and jotting down a few notes. They then explain their ideas to each another in pairs, merging them or selecting the most promising. Finally, a few pairs present their ideas to the whole group.

Think-pair-share works because it forces people to externalize their cognition (Section 3.1). It also gives them a chance to spot and resolve gaps or contradictions in their ideas *before* exposing them to a larger group, which can make less extroverted learners a little less nervous about appearing foolish.

MORNING, NOON, AND NIGHT

[Smar2018] found that learners do less well if their classes and other work are scheduled at times that don't line up with their natural body clocks, i.e., that if a morning person takes night classes or vice versa, their grades suffer. It's usually not possible to accommodate this in small groups, but larger ones should try to stagger start times for parallel sessions. This can also help people juggling childcare responsibilities and other constraints, and reduce the length of lineups at coffee breaks and for washrooms.

HUMOR

Humor should be used sparingly when teaching: most jokes are less funny when written down and become even less funny with each re-reading. Being spontaneously funny while teaching usually works better but can easily go wrong: what's a joke to your circle of friends may turn out to be a serious political issue to your audience. If you do make jokes when teaching, don't make them at the expense of any group, or of any individual except possibly yourself.

9.12   LIMIT INNOVATION

Each of the techniques presented in this chapter will make your classes better, but you shouldn't try to adopt them all at once. The reason is that every new practice increases *your* cognitive load as well as your learners', since you are all now trying to learn a new way to learn as well as the lesson's subject matter. If you are working with a group repeatedly, you can introduce one new technique every few lessons; if you only have them for a one-day workshop, it's best to pick just one method they haven't seen before and get them comfortable with that.

9.13   EXERCISES

CREATE A QUESTIONNAIRE (INDIVIDUAL/20)

Using the questionnaire in Section F.5 as a template, create a short questionnaire you could give learners before teaching a class of your own. What do you most want to know about their background, and how can both parties be sure they agree on what level of understanding you're asking about?

ONE OF YOUR OWN (WHOLE CLASS/15)

Think of one teaching practice that hasn't been described so far. Present your idea to a partner, listen to theirs, and select one to present to the group as a whole. (This exercise is an example of think-pair-share.)

MAY I DRIVE? (PAIRS/10)

Swap computers with a partner (preferably one who uses a different operating system than you) and work through a simple programming exercise. How frustrating is it? How much insight does it give you into what novices have to go through all the time?

PAIRING (PAIRS/15)

Watch this video[9] of pair programming and then practice doing it with a partner. Remember to switch roles between driver and navigator every few minutes. How long does it take you to fall into a working rhythm?

COMPARE NOTES (SMALL GROUPS/15)

Form groups of 3–4 people and compare the notes you have taken on this chapter. What did you think was noteworthy that your peers missed and vice versa? What did you understand differently?

CREDIBILITY (INDIVIDUAL/15)

[Fink2013] describes three things that make teachers credible in their learners' eyes:

Competence: knowledge of the subject as shown by the ability to explain complex ideas or reference the work of others.

Trustworthiness: having the learners' best interests in mind. This can be shown by giving individualized feedback, offering a rational explanation for grading decisions, and treating all learners the same.

Dynamism: excitement about the subject (Chapter 8).

Describe one thing you do when teaching that fits into each category, and then describe one thing you don't do but should.

MEASURING EFFECTIVENESS (INDIVIDUAL/15)

[Kirk1994] defines four levels at which to evaluate training:

Reaction: how did the learners feel about the training?

Learning: how much did they actually learn?

Behavior: how much have they changed their behavior as a result?

Results: how have those changes in behavior affected their output or the output of their group?

What are you doing at each level to evaluate what and how you teach? What could you do that you're not doing?

---

[9]https://www.youtube.com/watch?v=vgkahOzFH2Q

OBJECTIONS AND COUNTER-OBJECTIONS (THINK-PAIR-SHARE/15)

You have decided not to ask your learners if your class was useful because you know there is no correlation between their answers and how much they actually learn (Section 7.1). Instead, you have put forward four proposals, each of which your colleagues have shot down:

See if they recommend the class to friends. Why would this be any more meaningful than asking them how they feel about the class?

Give them an exam at the end. But how much learners know at the end of the day is a poor predictor of how much they will remember two or three months later, and any kind of final exam will make the class a lot more stressful.

Give them an exam two or three months later. That's practically impossible with free-range learners, and the people who didn't get anything out of the workshop are probably less likely to take part in follow-up, so feedback gathered this way will be skewed.

See if they keep using what they learned. Installing spyware on learners' computers is frowned upon, so how will this be implemented?

Working on your own, come up with answers to these objections, then share your responses with a partner and discuss the approaches you have come up with. When you are done, share your favored approach with the class.

10 Motivation and Demotivation

Learners need encouragement to step out into unfamiliar terrain, so this chapter discusses ways teachers can motivate them. More importantly, it talks about how teachers can *demotivate* them and how to avoid doing that.

Our starting point is the difference between **extrinsic motivation**, which we feel when we do something to avoid punishment or earn a reward, and **intrinsic motivation**, which is what we feel when we find something personally fulfilling. Both affect most situations—for example, people teach because they enjoy it and because they get paid—but we learn best when we are intrinsically motivated [Wlod2017]. According to self-determination theory[1], the three drivers of intrinsic motivation are:

Competence: the feeling that you know what you're doing.

Autonomy: the feeling of being in control of your own destiny.

Relatedness: the feeling of being connected to others.

A well-designed lesson encourages all three. For example, a programming exercise can let learners practice the tools they need to use to solve a larger problem (competence), let them tackle the parts of that problem in whatever order they want (autonomy), and allow them to talk to their peers (relatedness).

The Problem of Grades
I've never had an audience in my life. My audience is a rubric.
– quoted by Matt Tierney[2]

Grades and the way they distort learning are often used as an example of extrinsic motivation, but as [Mill2016a] observes, they aren't going to go away any time soon, so it's pointless to try to build a system that ignores them. Instead, [Lang2013] explores how courses that emphasize grades can incentivize learners to cheat and offers some tips on how to diminish this effect, while [Covi2017] looks at the larger problem of balancing intrinsic and extrinsic motivation in institutional education, and the constructive alignment[3] approach advocated in [Bigg2011] seeks to bring learning activities and learning outcomes into line with each other.

---

[1] https://en.wikipedia.org/wiki/Self-determination_theory

[2] https://twitter.com/figuralities/status/987330064571387906

[3] https://en.wikipedia.org/wiki/Constructive_alignment

[Ambr2010] contains a list of evidence-based methods to motivate learners. None of them are surprising—it's hard to imagine someone saying that we *shouldn't* identify and reward what we value—but it's useful to check lessons to make sure they are doing at least a few of these things. One strategy I particularly like is to have learners who struggled but succeeded come in and tell their stories to the rest of the class. Learners are far more likely to believe stories from people like themselves [Mill2016a], and people who have been through your course will always have advice you would never have thought of.

Not Just for Learners

Discussions of motivation in education often overlook the need to motivate the teacher. Learners respond to a teacher's enthusiasm, and teachers (particularly volunteers) need to care about a topic in order to keep teaching it. This is another powerful reason to co-teach (Section 9.3): just as having a running partner makes it more likely that you'll keep running, having a teaching partner helps get you up and going on those days when you have a cold and the projector bulb has burned out and nobody knows where to find a replacement and seriously, are they doing construction again?

Teachers can do other positive things as well. [Bark2014] found three things that drove retention for all learners: meaningful assignments, faculty interaction with learners, and learner collaboration on assignments. Pace and workload relative to expectations were also significant drivers, but primarily for male learners. Things that *didn't* drive retention were interactions with teaching assistants and interactions with peers in extracurricular activities. These results seem obvious, but the reverse would seem obvious too: if the study had found that extracurricular activities did drive retention, we would also think that made sense. Noticeably, two of the four retention drivers (faculty interaction and learner collaboration) take extra effort to replicate online (Chapter 11).

10.1 AUTHENTIC TASKS

As Dylan Wiliam points out in [Hend2017], motivation doesn't always lead to achievement, but achievement almost always leads to motivation: learners' success motivates them far more than being told how wonderful they are. We can use this idea in teaching by creating a grid whose axes are "mean time to master" and "usefulness once mastered" (Figure 10.1).

Things that are quick to master and immediately useful should be taught first, even if they aren't considered fundamental by people who are already competent practitioners, because a few early wins will build learners' confidence in themselves and their teacher. Conversely, things that are hard to learn and aren't useful to your learners at their current stage of development should be skipped entirely, while topics along the diagonal need to be weighed against each other.

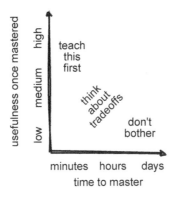

Figure 10.1: What to teach

Useful to Whom?

*If someone wants to build websites, foundational computer science con-
cepts like recursion and computability may inhabit the lower right corner of this grid. That doesn't mean they aren't worth learning, but if our
aim is to motivate people, they can and should be deferred. Conversely,
a senior who is taking a programming class to stimulate their mind may
prefer exploring these big ideas to doing anything practical. When you
are making up your grid, you should do it with your learner personas in
mind (Section 6.1). If topics wind up in very different places for different
personas, you should think about creating different courses.*

A well-studied instance of prioritizing what's useful without sacrificing
what's fundamental is the media computation approach developed at Georgia
Tech [Guzd2013]. Instead of printing "hello world" or summing the first ten integers, a learner's first program might open an image, resize it to create a thumbnail,
and save the result. This is an **authentic task**, i.e., something that learners believe
they would actually do in real life. It also has a **tangible artifact**: if the image comes
out the wrong size, learners have something in hand that can guide their debugging.
[Lee2013] describes an adaption of this approach from Python to MATLAB, while
others are building similar courses around data science, image processing, and biology [Dahl2018; Meys2018; Ritz2018].

There will always be tension between giving learners authentic problems and
exercising the individual skills they need to solve those problems: after all, programmers don't answer multiple choice questions on the job any more than musicians
play scales over and over in front of an audience. Finding the balance is hard, but a
first step is to take out anything arbitrary or meaningless. For example, programming
examples shouldn't use variables called foo and bar, and if you're going to have
learners sort a list, make it a list of songs rather than strings like "aaa" and "bbb".

10.2 DEMOTIVATION

Women aren't leaving computing because they don't know what it's like;
they're leaving because they *do* know.
— variously attributed

If you are teaching in a free-range setting, your learners are probably volunteers,
and probably want to be in your classroom. Motivating them is therefore less of a
concern than not demotivating them. Unfortunately, you can easily demotivate people by accident. For example, [Cher2009] reported four studies showing that subtle environmental clues have a measurable difference on the interest that people
of different genders have in computing: changing objects in a Computer Science
classroom from those considered stereotypical of computer science (e.g., Star Trek
posters and video games) to objects not considered stereotypical (e.g., nature posters
and phone books) boosted female undergraduates' interest to the level of their male
peers. Similarly, [Gauc2011] reports a trio of studies showing that gendered wording
commonly employed in job recruitment materials can maintain gender inequality in
traditionally male-dominated occupations.

There are three main demotivators for adult learners:

Unpredictability demotivates people because if there's no reliable connection
between what they do and what outcome they achieve, there's no reason for them
to try to do anything.

Indifference demotivates because learners who believe that the teacher or educational system doesn't care about them or the material won't care about it either.

Unfairness demotivates people who are disadvantaged for obvious reasons. What's
surprising is that it also demotivates people who benefit from unfairness: consciously or unconsciously, they worry that they will some day find themselves in
the disadvantaged group [Wilk2011].

In extreme situations, learners may develop **learned helplessness**: when repeatedly subjected to negative feedback in a situation that they can't change, they may
learn not to even try to change the things they could.

One of the fastest and surest ways to demotivate learners is to use language that
suggests that some people are natural programmers and others aren't. Guzdial has
called this the biggest myth about teaching computer science[4], and [Pati2016] backed
this up by showing that people see evidence for a "geek gene" where none exists.
They analyzed grade distributions from 778 university courses and found that only
5.8% showed signs of being multimodal: i.e., only one class in twenty showed signs
of having two distinct populations of learners. They then showed 53 Computer Science professors histograms of ambiguous grade distributions; those who believed
that some people are innately predisposed to be better at Computer Science were
more likely to see them as bimodal than those who didn't.

----

[4]https://cacm.acm.org/blogs/blog-cacm/189498-top-10-myths-about-teaching-computer-science/fulltext

These beliefs matter because teachers act on them [Brop1983]. If a teacher believes that a learner is likely to do well they naturally (often unconsciously) focus on that learner, who then fulfills the teacher's expectations because of the increased attention, which in turn appears to confirm the teacher's belief. Sadly, there is little sign that mere evidence of the kind presented in [Pati2016] is enough to break this vicious cycle. . .

Here are a few other specific things that will demotivate your learners:

A holier-than-thou or contemptuous attitude from a teacher or a fellow learner.

Telling them that their existing skills are rubbish. Unix users sneer at Windows, programmers of all kinds make jokes about Excel, and no matter what web application framework you already know, some programmer will tell you that it's out of date. Learners have often invested a lot of time and effort into acquiring the skills they have; disparaging them is a good way to guarantee that they won't listen to anything else you have to say.

Diving into complex or detailed technical discussion with the most advanced learners in the class.

Pretending that you know more than you do. Learners will trust you more if you are frank about the limitations of your knowledge, and will be more likely to ask questions and seek help.

Using the J word ("just") or feigning surprise. As discussed in Chapter 3, saying things like "I can't believe you don't know X" or "you've never heard of Y?" signals to the learner that the teacher thinks their problem is trivial and that they must be stupid for not being able to figure it out.

Software installation headaches. People's first contact with programming or with new programming tools is often demoralizing, and believing that something is hard to learn is a self-fulfilling prophecy. It isn't just the time it takes to get set up or the feeling that it's unfair to have to debug something that depends on precisely the knowledge they don't yet have. The real problem is that every such failure reinforces their belief that they would have a better chance of making next Thursday's deadline if they kept doing things the way they always have.

It is even easier to demotivate people online than in person, but there are now evidence-based strategies for dealing with this. [Ford2016] found that five barriers to contribution on Stack Overflow[5] are seen as significantly more problematic by women than men: lack of awareness of site features, feeling unqualified to answer questions, intimidating community size, discomfort interacting with or relying on strangers, and the feeling that searching for things online wasn't "real work." Fear of negative feedback didn't quite make this list, but would have been the next one added if the authors weren't quite so strict about their statistical cutoffs. All of these factors can and should be addressed in both in-person and online settings using methods like those in Section 10.4, and doing so improves outcomes for everyone [Sved2016].

---

[5]https://stackoverflow.com/

Productive Failure and Privilege

*Some recent work has explored **productive failure**, where learners are deliberately given problems that can't be solved with the knowledge they have and have to go out and acquire new information in order to make progress [Kapu2016]. Productive failure is superficially reminiscent of tech's "fail fast, fail often" mantra, but the latter is more a sign of privilege than of understanding. People can only afford to celebrate failure if they're sure they'll get a chance to try again; many of your learners, and many people from marginalized or underprivileged groups, can't be sure of that, and assuming that failure is an option is a great way to demotivate them.*

IMPOSTOR SYNDROME

Impostor syndrome is the belief that your achievements are lucky flukes and an accompanying fear that someone will finally figure this out. It is common among high achievers who undertake publicly visible work, but disproportionately affects members of under represented groups: as discussed in Section 7.1, [Wilc2018] found that female learners with prior exposure to computing outperformed their male peers in all areas in introductory programming courses but were consistently less confident in their abilities, in part because society keeps signaling in subtle and not-so-subtle ways that they don't really belong.

Traditional classrooms can fuel impostor syndrome. Schoolwork is frequently undertaken alone or in small groups, but the results are shared and criticized publicly. As a result, we rarely see how others struggle to finish their work, which can feed the belief that everyone else finds this easy. Members of underrepresented groups who already feel additional pressure to prove themselves may be particularly affected.

The Ada Initiative has created some guidelines[6] for fighting your own impostor syndrome, which include:

Talk about the issue with people you trust. When you hear from others that impostor syndrome is a common problem, it becomes harder to believe your feelings of being a fraud are real.

Go to an in-person impostor syndrome session. There's nothing like being in a room full of people you respect and discovering that 90% of them have impostor syndrome.

Watch your words, because they influence how you think. Saying things like, "I'm not an expert in this, but..." detracts from the knowledge you actually possess.

Teach others about your field. You will gain confidence in your own knowledge and skill and help others avoid some impostor syndrome shoals.

---

[6]https://www.usenix.org/blog/impostor-syndrome-proof-yourself-and-your-community

Ask questions. Asking questions can be intimidating if you think you should know the answer, but getting answers eliminates the extended agony of uncertainty and fear of failure.

Build alliances. Reassure and build up your friends, who will reassure and build you up in return. (If they don't, you might want to think about finding new friends...)

Own your accomplishments. Keep actively recording and reviewing what you have done, what you have built, and what successes you've had.

As a teacher, you can help people with their impostor syndrome by sharing stories of mistakes that you have made or things you struggled to learn. This reassures the class that it's OK to find topics hard. Being open with the group also builds trust and gives them confidence to ask questions. (Live coding is great for this: as noted in Section 8.1, your typos show your class that you're human.) Frequent formative assessments help as well, particularly if learners see you adjusting what you teach or how quickly you go based on their outcomes.

MINDSET AND STEREOTYPE THREAT

Carol Dweck and others have studied the differences of **fixed mindset** and **growth mindset** on learning outcomes. If people believe that competence in some area is intrinsic (i.e., that you either "have the gene" for it or you don't), *everyone* does worse, including the supposedly advantaged. The reason is that if someone doesn't do well at first, they assume that they lack that aptitude, which biases their future performance. On the other hand, if people believe that a skill is learned and can be improved, they do better on average.

There are concerns[7] that growth mindset has been oversold, or that it is much more difficult to translate research about it into practice than its more enthusiastic advocates have implied [Sisk2018]. However, it does appear that learners with low socioeconomic status or who are academically at risk might benefit from mindset interventions.

Another widely discussed effect is **stereotype threat** [Stee2011]. Reminding people of negative stereotypes, even in subtle ways, can make them anxious about the risk of confirming those stereotypes, which in turn can reduce their performance. Again, there are some concerns about the replicability of key studies[8], and the issue is further clouded by the fact that the term has been used in many ways [Shap2007], but no one would argue that mentioning stereotypes in class will help learners.

---

[7] https://educhatter.wordpress.com/2017/03/26/growth-mindset-is-the-theory-flawed-or-has-gm-been-debased-in-the-classroom/

[8] https://www.psychologytoday.com/blog/rabble-rouser/201512/is-stereotype-threat-overcooked-overstated-and-oversold

10.3 ACCESSIBILITY

Putting lessons and exercises out of someone's reach is about as demotivating as it gets, and it's very easy to do this inadvertently. For example, the first online programming lessons I wrote had a transcript of the narration beside the slides, but didn't include the actual source code: that was in screenshots of PowerPoint slides. Someone using a screen reader[9] could therefore hear what was being said about the program, but wouldn't know what the program actually was. It isn't always feasible to accommodate every learner's needs, but adding description captions to images and making navigation controls accessible to people who can't use a mouse can make a big difference.

Curb Cuts

Making material accessible helps everyone, not just people facing challenges. Curb cuts[10]—the small sloped ramps joining a sidewalk to the street—were originally created to make it easier for the physically disabled to move around, but proved to be equally helpful to people with strollers and grocery carts. Similarly, captioning images doesn't just help the visually impaired: it also makes images easier for search engines to find and index.

The first and most important step in making lessons accessible is to involve people with disabilities in decision making: the slogan *nihil de nobis, sine nobis*[11] (literally, "nothing for us without us") predates accessibility rights, but is always the right place to start. A few specific recommendations are:

Find out what you need to do. Each of these posters[12] offers do's and don'ts for people on the autistic spectrum, users of screen readers, and people with low vision, physical or motor disabilities, hearing exercises, and dyslexia.

Don't do everything at once. The enhancements described in the previous point can seem pretty daunting, so make one change at a time.

Do the easy things first. Font size, using a clip-on microphone so that people can hear you more easily, and checking your color choices are good places to start.

Know how well you're doing. Sites like WebAIM[13] allow you to check how accessible your online materials are to visually impaired users.

[Coom2012; Burg2015] are good guides to visual design for accessibility. Their recommendations include:

---

[9]https://en.wikipedia.org/wiki/Screen_reader

[10]https://en.wikipedia.org/wiki/Curb_cut

[11]https://en.wikipedia.org/wiki/Nothing_About_Us_Without_Us

[12]https://accessibility.blog.gov.uk/2016/09/02/dos-and-donts-on-designing-for-accessibility/

[13]http://webaim.org/

Format documents with actual headings and other landmarks rather than just changing font sizes and styles.

Avoid using color alone to convey meaning in text or graphics. Instead, use color plus different cross-hatching patterns (which also makes material understandable when printed in black and white).

Remove unnecessary elements rather than just making them invisible, because screen readers will still often say them aloud.

Allow self-pacing and repetition for people with reading or hearing issues.

Include narration of on-screen action in videos (and talk while you type when live coding).

SPOONS

In 2003, Christine Miserandino started using spoons[14] as a way to explain what it's like to live with chronic illness. Healthy people start each day with an unlimited supply of spoons, but people with lupus or other debilitating conditions only have a few, and everything they do costs them one. Getting out of bed? That's a spoon. Making a meal? That's another spoon, and pretty soon, you've run out.

> You cannot simply just throw clothes on when you are sick... If my hands hurt that day buttons are out of the question. If I have bruises that day, I need to wear long sleeves, and if I have a fever I need a sweater to stay warm and so on. If my hair is falling out I need to spend more time to look presentable, and then you need to factor in another 5 minutes for feeling badly that it took you 2 hours to do all this.

As Elizabeth Patitsas has argued[15], people who have a lot of spoons can accumulate more, but people whose supply is limited may struggle to get ahead. When designing classes and exercises, remember that some of your learners may have physical or mental obstacles that aren't obvious. When in doubt, ask: they almost certainly have more experience with what works and what doesn't than anyone else.

10.4   INCLUSIVITY

Inclusivity is a policy of including people who might otherwise be excluded or marginalized. In computing, it means making a positive effort to be more welcoming to women, under represented racial or ethnic groups, people with various sexual orientations, the elderly, those facing physical challenges, the formerly incarcerated, the economically disadvantaged, and everyone else who doesn't fit Silicon Valley's affluent white/Asian male demographic. Figure 10.2 (from NPR[16]) graphically illustrates the effects of computing's exclusionary culture on women.

---

[14]https://butyoudontlooksick.com/articles/written-by-christine/the-spoon-theory/

[15]https://patitsas.blogspot.com/2018/03/spoons-are-form-of-capital.html

[16]https://www.npr.org/sections/money/2014/10/21/357629765/when-women-stopped-coding

What Happened To Women In Computer Science?

% Of Women Majors, By Field

■ Medical School   ■ Law School   ▓ Physical Sciences   ■ Computer science

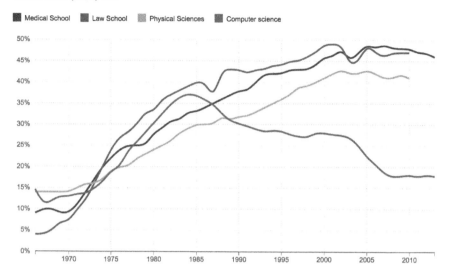

Figure 10.2: Female computer science majors in the US

[Lee2017] is a brief, practical guide to doing that with references to the research literature. The practices it describes help learners who belong to one or more marginalized or excluded groups, but help motivate everyone else as well. They are phrased in terms of term-long courses, but many can be applied in workshops and other free-range settings:

Ask learners to email you before the workshop to explain how they believe the training could help them achieve their goals.

Review your notes to make sure they are free from gendered pronouns, include culturally diverse names, etc.

Emphasize that what matters is the rate at which they are learning, not the advantages or disadvantages they had when they started.

Encourage pair programming, but demonstrate it first so that learners understand the roles of driver and navigator.

Actively mitigate behavior that some learners may find intimidating, e.g., use of jargon or "questions" that are actually asked to display knowledge.

One way to support learners from marginalized groups is to have people sign up for workshops in groups rather than individually. That way, everyone in the room knows in advance that they will be with people they trust, which increases the chances of them actually coming. It also helps after the workshop: if people come with their friends or colleagues, they can work together to use what they've learned.

More fundamentally, lesson authors need to take everyone's entire situation into account. For example, [DiSa2014a] found that 65% of male African-American participants in a game testing program went on to study computing, in part because the gaming aspect of the program was something their peers respected. [Lach2018] explored two general strategies for creating inclusive content and the risks associated with them:

Community representation highlights learners' social identities, histories, and community networks using after-school mentors or role models from learners' neighborhoods, or activities that use community narratives and histories as a foundation for a computing project. The major risk with this approach is shallowness, e.g., using computers to build slideshows rather than do any real computing.

Computational integration incorporates ideas from the learner's community, such as reproducing indigenous graphic designs in a visual programming environment. The major risk here is cultural appropriation, e.g., using practices without acknowledging origins.

If in doubt, ask your learners and members of the community what they think you ought to do. We return to this in Chapter 13.

Conduct as Accessibility

We said in Section 9.1 that classes should enforce a Code of Conduct like the one in Appendix B. This is a form of accessibility: while closed captions make video accessible to people with hearing disabilities, a Code of Conduct makes lessons accessible to people who would otherwise be marginalized.

MOVING PAST THE DEFICIT MODEL

Depending on whose numbers you trust, only 12–18% of people getting computer science degrees are women, which is less than half the percentage seen in the mid-1980s (Figure 10.3). And western countries are the odd ones for having such a low percentage of women in computing: women are still often 30–40% of computer science students elsewhere [Galp2002; Varm2015].

Since it's unlikely that women have changed drastically in the last 30 years, we have to look for structural causes to understand what's gone wrong and how to fix it. One explanation is the way that home computers were marketed as "boys' toys" starting in the 1980s [Marg2003]; another is the way that computer science departments responded to explosive growth in enrollment in the 1980s and again in the 2000s by changing admission requirements [Robe2017]. None of these factors may seem dramatic to people who aren't affected by them, but they act like the steady drip of water on a stone: over time, they erode motivation, and with it, participation.

The first and most important step toward fixing this is to stop thinking in terms of a "leaky pipeline" [Mill2015]. More generally, we need to move past a **deficit model**, i.e., to stop thinking that the members of under represented groups lack something

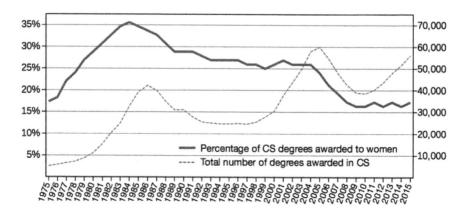

Figure 10.3: Degrees awarded and female enrollment [Robe2017]

and are therefore responsible for not getting ahead. Believing that puts the burden on people who already have to do extra work to overcome structural inequities and (not coincidentally) gives those who benefit from the current arrangements an excuse not to look at themselves too closely.

Rewriting History

[Abba2012] describes the careers and accomplishments of the women who shaped the early history of computing, but have all too often been written out of it; [Ensm2003; Ensm2012] describes how programming was turned from a female into a male profession in the 1960s, while [Hick2018] looks at how Britain lost its early dominance in computing by systematically discriminating against its most qualified workers: women. (See [Milt2018] for a review of all three books.) Discussing this history makes some men in computing very uncomfortable; in my opinion, that's a good reason to do it.

Misogyny in video games, the use of "cultural fit" in hiring to excuse conscious or unconscious bias, a culture of silence around harassment, and the growing inequality in society that produces preparatory privilege (Section 9.5) are not any one person's fault, but fixing them is everyone's responsibility. As a teacher, you have more power than most; this workshop[17] has excellent practical advice on how to be a good ally, and its advice is probably more important than anything this book teaches you about teaching.

---

[17] https://frameshiftconsulting.com/ally-skills-workshop/

10.5 EXERCISES

AUTHENTIC TASKS (PAIRS/15)

1. In pairs, list half a dozen things you did this week that use the skills you teach.
2. Place your items on a 2x2 grid of "time to master" and "usefulness". Where do you agree and disagree?

CORE NEEDS (WHOLE CLASS/10)

Paloma Medina identifies six core needs[18] for people at work: belonging, improvement (i.e., making progress), choice, equality, predictability, and significance. After reading her description of these, order them from most to least significant for you personally, then compare rankings with your peers. How do you think your rankings compare with those of your learners?

IMPLEMENT ONE STRATEGY FOR INCLUSIVITY (INDIVIDUAL/5)

Pick one activity or change in practice from [Lee2017] that you would like to work on. Put a reminder in your calendar three months in the future to ask yourself whether you have done something about it.

AFTER THE FACT (THINK-PAIR-SHARE/20)

1. Think back to a course that you took in the past and identify one thing the teacher did that demotivated you. Make notes about what could have been done afterward to correct the situation.
2. Pair up with your neighbor and compare stories, then add your comments to a set of notes shared by the whole class.
3. Review the comments in the shared notes as a group. Highlight and discuss a few of the things that could have been done differently.
4. Do you think that doing this will help you handle situations like these in the future?

WALK THE ROUTE (WHOLE CLASS/15)

Find the nearest public transportation drop-off point to your building and walk from there to your office and then to the nearest washroom, making notes about things you think would be difficult for someone with mobility issues. Now borrow a wheelchair and repeat the journey. How complete was your list of exercises? And did you notice that the first sentence in this exercise assumed you could actually walk?

---

[18]https://www.palomamedina.com/biceps

WHO DECIDES? (WHOLE CLASS/15)

In [Litt2004], Kenneth Wesson wrote, "If poor inner-city children consistently outscored children from wealthy suburban homes on standardized tests, is anyone naive enough to believe that we would still insist on using these tests as indicators of success?" Read this article[19] by Cameron Cottrill, and then describe an example from your own experience of "objective" assessments that reinforced the status quo.

COMMON STEREOTYPES (PAIRS/10)

Some people still say, "It's so simple that even your grandmother could use it." In pairs, list two or three other phrases that reinforce stereotypes about computing.

NOT BEING A JERK (INDIVIDUAL/15)

This short article[20] by Gary Bernhardt rewrites an unnecessarily hostile message to be less rude. Using it as a model, find something unpleasant on Stack Overflow[21] or some other public discussion forum and rewrite it to be more inclusive.

SAVING FACE (INDIVIDUAL/10)

Would any of your hoped-for learners be embarrassed to admit that they don't already know some of the things you want to teach? If so, how can you help them save face?

CHILDHOOD TOYS (WHOLE CLASS/15)

[Cutt2017] surveyed adult computer users about their childhood activities and found that the strongest correlation between confidence and computer use were based on reading on one's own and playing with construction toys like Lego that do not having moving parts. Survey the class and see what other activities people engaged in, then search for these activities online. How strongly gendered are descriptions and advertising for them? What effect do you think this has?

LESSON ACCESSIBILITY (PAIRS/30)

In pairs, choose a lesson whose materials are available online and independently rank it according to the do's and don'ts in these posters[22]. Where did you and your partner agree? Where did you disagree? How well did the lesson do for each of the six categories of user?

---

[19] https://mobile.nytimes.com/2016/04/10/upshot/why-talented-black-and-hispanic-students-can-go-undiscovered.html

[20] https://www.destroyallsoftware.com/blog/2018/a-case-study-in-not-being-a-jerk-in-open-source

[21] https://stackoverflow.com/

[22] https://accessibility.blog.gov.uk/2016/09/02/dos-and-donts-on-designing-for-accessibility/

TRACING THE CYCLE (SMALL GROUPS/15)

[Coco2018] traces a depressingly common pattern in which good intentions are undermined by an organization's leadership being unwilling to actually change. Working in groups of 4–6, write brief texts or emails that you imagine each of the parties involved would send to the other at each stage in this cycle.

WHAT'S THE WORST THING THAT COULD HAPPEN? (SMALL GROUPS/5)

Over the years, I have had a projector catch fire, a student go into labor, and a fight break out in class. I've fallen off stage twice, fallen asleep in one of my own lectures, and had many jokes fall flat. In small groups, make up a list of the worst things that have happened to you while you were teaching, then share with the class. Keep the list to remind yourself later that no matter how bad class was, at least none of *that* happened.

REVIEW

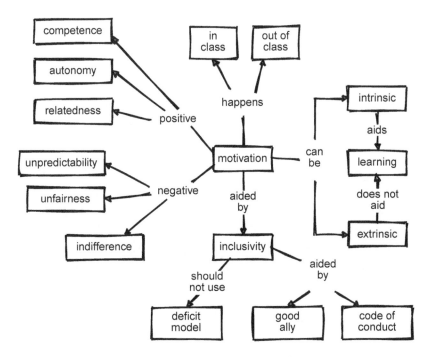

Figure 10.4: Concepts: Motivation

11 Teaching Online

If you use robots to teach, you teach people to be robots.
— variously attributed

Technology has changed teaching and learning many times. Before blackboards were introduced into schools in the early 1800s, there was no way for teachers to share an improvised example, diagram, or exercise with an entire class at once. Cheap, reliable, easy to use, and flexible, blackboards enabled teachers to do things quickly and at a scale that they had only been able to do slowly and piecemeal before. Similarly, hand-held video cameras revolutionized athletics training, just as tape recorders revolutionized music instruction a decade earlier.

Many of the people pushing the Internet into classrooms don't know this history, and don't realize that theirs is just the latest in a long series of attempts[1] to use machines to teach [Watt2014]. From the printing press through radio and television to desktop computers and mobile devices, every new way to share knowledge has produced a wave of aggressive optimists who believe that education is broken and that technology can fix it. However, ed tech's loudest advocates have often known less about "ed" than "tech," and behind their rhetoric, many have been driven more by the prospect of profit than by the desire to empower learners.

Today's debate is often muddied by confusing "online" with "automated." Run well, a dozen people working through a problem in a video chat feels like any other small-group discussion. Conversely, a squad of teaching assistants grading hundreds of papers against an inflexible rubric might as well be a collection of Perl scripts. This chapter therefore starts by looking at fully automated online instruction using recorded videos and automatically graded exercises, then explores some alternative hybrid models.

11.1 MOOCS

The highest-profile effort to reinvent education using the Internet is the **Massive Open Online Course**, or MOOC. The term was invented by David Cormier in 2008 to describe a course organized by George Siemens and Stephen Downes. That course was based on a **connectivist** view of learning, which holds that knowledge is distributed and that learning is the process of finding, creating, and pruning connections.

The term "MOOC" was quickly co-opted by creators of courses more closely resembled the hub-and-spoke model of a traditional classroom, with the teacher at the center defining goals and the learners seen as recipients or replicators of knowledge. Classes that use the original connectivist model are now sometimes referred to as

---

[1] http://teachingmachin.es/timeline.html

"cMOOCs," while classes that centralize control are called "xMOOCs." (The latter is also sometimes called a "MESS," for Massively Enhanced Sage on the Stage.)

Five years ago, you couldn't walk across a major university campus without hearing some talking about how MOOCs would revolutionize education, destroy it, or possibly both. MOOCs would give learners access to a wider range of courses and allow them to work when it was convenient for them rather than fitting their learning to someone else's schedule.

But MOOCs haven't been nearly as effective as their more enthusiastic proponents predicted [Ubel2017]. One reason is that recorded content is ineffective for many novices because it cannot clear up their individual misconceptions (Chapter 2): if they don't understand an explanation the first time around, there usually isn't a different one on offer. Another is that the automated assessment needed to put the "massive" in MOOC only works well at the lowest levels of Bloom's Taxonomy. It's also now clear that learners have to shoulder much more of the burden of staying focused in a MOOC, that the impersonality of working online can encourage uncivil behavior and demotivate people, and that "available to everyone" actually means "available to everyone affluent enough to have high-speed Internet and lots of free time."

[Marg2015] examined 76 MOOCs on various subjects and found that while the organization and presentation of material was good, the quality of lesson design was poor. Closer to home, [Kim2017] studied 30 popular online coding tutorials, and found that they largely taught the same content the same way: bottom-up, starting with low-level programming concepts and building up to high-level goals. Most required learners to write programs and provided some form of immediate feedback, but this feedback was typically very shallow. Few explained when and why concepts are useful (i.e., they didn't show how to transfer knowledge) or provided guidance for common errors, and other than rudimentary age-based differentiation, none personalized lessons based on prior coding experience or learner goals.

Personalized Learning

*Few terms have been used and abused in as many ways as **personalized learning**. To most ed tech proponents, it means dynamically adjusting the pace of lessons based on learner performance, so that if someone answers several questions in a row correctly, the computer will skip some of the subsequent questions.*

Doing this can produce modest improvements[2], but better is possible. For example, if many learners find a particular topic difficult, the teacher can prepare multiple alternative explanations of that point rather than accelerating a single path. That way, if one explanation doesn't resonate, others are available. However, this requires a lot more design work on the teacher's part, which may be why it hasn't proven popular. And even if it does work, the effects are likely to be much less than some of its advocates believe. A good teacher makes a difference of 0.1–0.15 standard deviations

---

[2]https://www.rand.org/pubs/research_briefs/RB9994.html

in end-of-year performance in grade school [Chet2014] (see this article[3]
for a brief summary). It's unrealistic to believe that any kind of automation
can outdo this any time soon.

So how *should* the Internet be used in teaching and learning tech skills? Its pros
and cons are:

Learners can access more lessons, more quickly, than ever before. Provided, of
course, that a search engine considers those lessons worth indexing, that their
Internet service provider and government don't block it, and that the truth isn't
drowned in a sea of attention-sapping disinformation.

Learners can access *better* lessons than ever before, unless they are being steered
toward second-rate material in order to redistribute wealth from the have-nots to
the haves [McMi2017]. It's also worth remembering that scarcity increases per-
ceived value, so as online education becomes cheaper, it will increasingly become
what everyone wants for someone else's children.

Learners can access far more people than ever before as well. But only if those
learners actually have access to the required technology, can afford to use it, and
aren't driven offline by harassment or marginalized because they don't conform to
the social norms of whichever group is talking loudest. In practice, most MOOC
users come from secure, affluent backgrounds [Hans2015].

Teachers can get far more detailed insight into how learners work. So long as
learners are doing things that are amenable to large-scale automated analysis and
either don't object to surveillance in the classroom or aren't powerful enough for
their objections to matter.

[Marg2015; Mill2016a; Nils2017] describe ways to accentuate the positives in the
list above while avoiding the negatives:

Make deadlines frequent and well-publicized, and enforce them so that learners
will get into a work rhythm.

Keep synchronous all-class activities like live lectures to a minimum so that people don't miss things because of scheduling conflicts.

Have learners contribute to collective knowledge, e.g., take notes together (Section 9.7), serve as classroom scribes, or contribute problems to shared problem
sets (Section 5.3).

Encourage or require learners to do some of their work in small groups that *do*
have synchronous online activities such as a weekly online discussion. This
helps learners stay engaged and motivated without creating too many scheduling headaches. (See Appendix E for some tips on how to make these discussions
fair and productive.)

---

[3] http://educationnext.org/in-schools-teacher-quality-matters-most-coleman/

Create, publicize, and enforce a code of conduct so that everyone can actually take part in online discussions (Section 9.1).

Use lots of short lesson episodes rather than a handful of lecture-length chunks in order to minimize cognitive load and provide lots of opportunities for formative assessment. This also helps with maintenance: if all of your videos are short, you can simply re-record any that need maintenance, which is often cheaper than trying to patch longer ones.

Use video to engage rather than instruct. Disabilities aside (Section 10.3), learners can read faster than you can talk. The exception to this rule is that video is actually the best way to teach people verbs (actions): short screencasts that show people how to use an editor, step through code in a debugger, and so on are more effective than screenshots with text.

Identify and clear up misconceptions early. If data shows that learners are struggling with some parts of a lesson, create alternative explanations of those points and extra exercises for them to practice on.

All of this has to be implemented somehow, which means that you need some kind of teaching platform. You can either use an all-in-one **learning management system** like Moodle[4] or Sakai[5], or assemble something yourself using Slack[6] or Zulip[7] for chat, Google Hangouts[8] or appear.in[9] for video conversations, and Word-Press[10], Google Docs[11], or any number of wiki systems for collaborative authoring. If you are just starting out, pick whatever is easiest to set up and administer and is most familiar to your learners. If faced with a choice, the second consideration is more important than the first: you're expecting people to learn a lot in your class, so it's only fair for you to learn how to drive the tools they're most comfortable with.

Assembling a platform for learning is necessary but not sufficient: if you want your learners to thrive, you need to create a community. Hundreds of books and presentations talk about how to do this, but most are based on their authors' personal experiences. [Krau2016] is a welcome exception: while it predates the accelerating descent of Twitter and Facebook into weaponized abuse and misinformation, most of its findings are still relevant. [Foge2005] is also full of useful tips about the communities of practice that learners may hope to join; we explore some of its ideas in Chapter 13.

---

[4] http://moodle.org

[5] https://www.sakaiproject.org/

[6] http://slack.com

[7] https://zulipchat.com/

[8] http://hangouts.google.com

[9] https://appear.in/

[10] https://wordpress.org/

[11] http://docs.google.com

Freedom To and Freedom From

Isaiah Berlin's 1958 essay "Two Concepts of Liberty[12]" made a distinction between positive liberty, which is the ability to actually do something, and negative liberty, which is the absence of rules saying that you can't do it. Online discussions usually offer negative liberty (nobody's stopping you from saying what you think) but not positive liberty (many people can't actually be heard). One way to address this is to introduce some kind of throttling, such as only allowing each learner to contribute one message per discussion thread per day. Doing this gives those with something to say a chance to say it, while clearing space for others to say things as well.

One other concern people have about teaching online is cheating. Day-to-day dishonesty is no more common in online classes than in face-to-face settings [Beck2014], but the temptation to have someone else write the final exam, and the difficulty of checking whether this happened, is one of the reasons educational institutions have been reluctant to offer credit for pure online classes. Remote exam proctoring is possible, but before investing in this, read [Lang2013]: it explores why and how learners cheat, and how courses can be structured to avoid giving them a reason to do so.

11.2   VIDEO

A prominent feature of most MOOCs is their use of recorded video lectures. These can be effective: as mentioned in Chapter 8, a teaching technique called Direct Instruction based on precise delivery of a well-designed script has repeatedly been shown to be effective [Stoc2018]. However, scripts for direct instruction have to be designed, tested, and refined very carefully, which is an investment that many MOOCs have been unwilling or unable to make. Making a small change to a web page or a slide deck only takes a few minutes; making even a small change to a short video takes an hour or more, so the cost to the teacher of acting on feedback can be unsupportable. And even when they're well made, videos have to be combined with activities to be beneficial: [Koed2015] estimated, "...the learning benefit from extra doing...to be more than six times that of extra watching or reading."

If you are teaching programming, you may use screencasts instead of slides, since they offer some of the same advantages as live coding (Section 8.1). [Chen2009] offers useful tips for creating and critiquing screencasts and other videos; Figure 11.1 (from [Chen2009]) reproduces the patterns that paper presents and the relationships between them. (It's also a good example of a concept map (Section 3.1).)

So what makes an instructional video effective? [Guo2014] measured engagement by looking at how long learners watched MOOC videos, and found that:

• Shorter videos are much more engaging—videos should be no more than six minutes long.

---

[12]https://en.wikipedia.org/wiki/Two_Concepts_of_Liberty

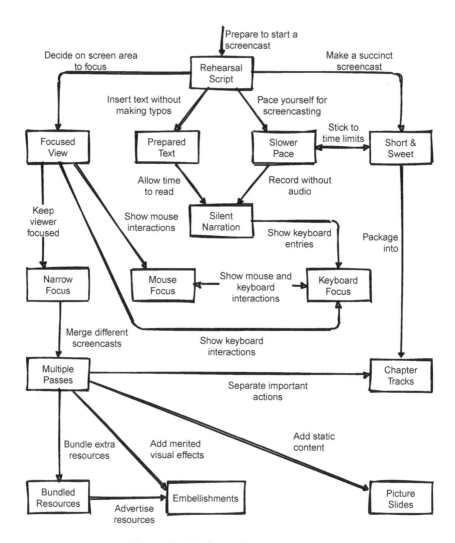

Figure 11.1: Patterns for screencasting

- A talking head superimposed on slides is more engaging than voice over slides alone.
- Videos that felt personal could be more engaging than high-quality studio recordings, so filming in informal settings could work better than professional studio work for lower cost.
- Drawing on a tablet is more engaging than PowerPoint slides or code screencasts, though it's not clear whether this is because of the motion and informality or because it reduces the amount of text on the screen.
- It's OK for teachers to speak fairly fast as long as they are enthusiastic.

One thing [Guo2014] didn't address is the chicken-and-egg problem: do learners find a certain kind of video engaging because they're used to it, so producing more videos of that kind will increase engagement simply because of a feedback loop? Or do these recommendations reflect some deeper cognitive processes? Another thing this paper didn't look at is learning outcomes: we know that learner evaluations of courses don't correlate with learning [Star2014; Uttl2017], and while it's plausible that learners won't learn from things they don't watch, it remains to be proven that they *do* learn from things they *do* watch.

I'm a Little Uncomfortable

[Guo2014]'s research was approved by a university research ethics board, the learners whose viewing habits were monitored almost certainly clicked "agree" on a terms of service agreement at some point, and I'm glad to have these insights. On the other hand, the word "privacy" didn't appear in the title or abstract of any *of the dozens of papers or posters at the conference where these results were presented. Given a choice, I'd rather not know how engaged learners are than foster ubiquitous surveillance in the classroom.*

There are many different ways to record video lessons; to find out which are most effective, [Mull2007a] assigned 364 first-year physics learners to online multimedia treatments of Newton's First and Second Laws in one of four styles:

Exposition: concise lecture-style presentation.

Extended Exposition: as above with additional interesting information.

Refutation: Exposition with common misconceptions explicitly stated and refuted.

Dialog: Learner-tutor discussion of the same material as in the Refutation.

Refutation and Dialog produced the greatest learning gains compared to Exposition; learners with low prior knowledge benefited most, and those with high prior knowledge were not disadvantaged. Again, this highlights the importance of directly addressing learners' misconceptions. Don't just tell people what *is*: tell them what *isn't* and why not.

11.3 HYBRID MODELS

Fully automated teaching is only one way to use the web in teaching. In practice, almost all learning in affluent societies has an online component today, either officially or through peer-to-peer back channels and surreptitious searches for answers to homework questions. Combining live and automated instruction allows teachers to use the strengths of both. In a traditional classroom, the teacher can answer questions immediately, but it takes days or weeks for learners to get feedback on their coding exercises. Online, it can take longer for a learner to get an answer, but they can get immediate feedback on their coding (at least for those kinds of exercises we can auto-grade).

Another difference is that online exercises have to be more detailed because they have to anticipate learners' questions. I find that in-person lessons start with the intersection of what everyone needs to know and expands on demand, while online lessons have to include the union of what everyone needs to know because the teacher isn't there to do the expanding.

In reality, the distinction between online and in-person is now less important for most people than the distinction between synchronous and asynchronous: do teachers and learners interact in real time, or is their communication spread out and interleaved with other activities? In-person will almost always be synchronous, but online is increasingly a mixture of both:

> I think that our grandchildren will probably regard the distinction we make between what we call the real world and what they think of as simply the world as the quaintest and most incomprehensible thing about us.
> — William Gibson

The most popular implementation of this blended future today is the **flipped classroom**, in which learners watch recorded lessons on their own and class time is used for discussion and working through problem sets. Originally described in [King1993], the idea was popularized as part of peer instruction (Section 9.2) and has been studied intensively over the past decade. For example, [Camp2016] compared learners who took an introductory computer science class online with those who took it in a flipped classroom. Completion of (unmarked) practice exercises correlated with exam scores for both, but the completion rate of rehearsal exercises by online learners was significantly lower than lecture attendance rates for in-person learners.

But if recordings are available, will learners still show up to class to do practice exercises? [Nord2017] examined the impact of recordings on both lecture attendance and learners' performance at different levels. In most cases the study found no negative consequences of making recordings available; in particular, learners didn't skip lectures when recordings are available (at least, not any more than they usually do). The benefits of providing recordings were greatest for learners early in their careers, but diminished as learners become more mature.

Another hybrid model brings online life into the classroom. Taking notes together is a first step (Section 9.7); pooling answers to multiple choice questions in real time using tools like Pear Deck[13] and Socrative[14] is another. If the class is small—say, a dozen to fifteen people—you can also have all of the learners join a video conference so that they can screenshare with the teacher. This allows them to show their work (or their problems) to the entire class without having to connect their laptop to the projector. Learners can also then use the chat in the video call to post questions for the teacher; in my experience, most of them will be answered by their fellow learners, and the teacher can handle the rest when they reach a natural break. This model helps level the playing field for remote learners: if someone isn't able to attend class for health reasons or because of family or work commitments, they can still take part on a nearly-equal basis if everyone is used to collaborating online in real time.

I have also delivered classes using real-time remote instruction, in which learners are co-located at 2–6 sites with helpers present while I taught via streaming video (Section C.1). This scales well, saves on travel costs, and allows the use of techniques like pair programming (Section 9.6). What *doesn't* work is having one group in person and one or more groups remotely: with the best will in the world, the local participants get far more attention.

11.4 ONLINE ENGAGEMENT

[Nuth2007] found that there are three overlapping worlds in every classroom: the public (what the teacher is saying and doing), the social (peer-to-peer interactions between learners), and the private (inside each learner's head). Of these, the most important is usually the social: learners pick up as much via cues from their peers as they do from formal instruction.

The key to making any form of online teaching effective is therefore to facilitate peer-to-peer interactions. To aid this, courses almost always have some kind of discussion forum. [Mill2016a] observed that learners use these in very different ways:

> ...procrastinators are particularly unlikely to participate in online discussion forums, and this reduced participation, in turn, is correlated with worse grades. A possible explanation for this correlation is that procrastinators are especially hesitant to join in once the discussion is under way, perhaps because they worry about being perceived as newcomers in an established conversation. This aversion to jump in late causes them to miss out on the important learning and motivation benefits of peer-to-peer interaction.

[Vell2017] analyzes discussion forum posts from 395 CS2 students at two universities by dividing them into four categories:

Active: request for help that does not display reasoning and doesn't display what the student has already tried or already knows.

---

[13]https://www.peardeck.com/

[14]https://socrative.com/

Constructive: reflect students' reasoning or attempts to construct a solution to the problem.

Logistical: course policies, schedules, assignment submission, etc.

Content clarification: request for additional information that doesn't reveal the student's own thinking.

They found that constructive and logistical questions dominated, and that constructive questions correlated with grades. They also found that students rarely ask more than one active question in a course, and that these *don't* correlate with grades. While this is disappointing, knowing it helps set teachers' expectations: while we might all want our courses to have lively online communities, we have to accept that most won't, or that most learner-to-learner discussion will take place through channels that they are already using that we may not be part of.

Co-opetition

[Gull2004] describes an online coding contest that combines collaboration and competition. The contest starts when a problem description is posted along with a correct but inefficient solution. When it ends, the winner is the person who has made the greatest overall contribution to improving the performance of the overall solution. All submissions are in the open, so that participants can see one another's work and borrow ideas from each other. As the paper shows, the final solution is almost always a hybrid borrowing ideas from many people.

[Batt2018] described a small-scale variation of this in an introductory computing class. In stage one, each learner submitted a programming project individually. In stage two, learners were paired to create an improved solution to the same problem. The assessment indicates that two-stage projects tend to improve learners' understanding and that they enjoyed the process. Projects like these not only improve engagement, they also give participants more experience building on someone else's code.

Discussion isn't the only way to get learners to work together online. [Pare2008] and [Kulk2013] report experiments in which learners grade each other's work, and the grades they assign are then compared with the grades given by graduate-level teaching assistants or other experts. Both found that learner-assigned grades agreed with expert-assigned grades as often as the experts' grades agreed with each other, and that a few simple steps (such as filtering out obviously unconsidered responses or structuring rubrics) decreased disagreement even further. And as discussed in Section 5.3, collusion and bias are *not* significant factors in peer grading.

Trust, but Educate

The most common way to measure the validity of feedback is to compare learners' grades to experts' grades, but calibrated peer review (Section 5.3) can be equally effective. Before asking learners to grade each others' work, they are asked to grade samples and compare their results with the grades assigned by the teacher. Once the two align, the learner is

*allowed to start giving grades to peers. Given that critical reading is an
effective way to learn, this result may point to a future in which learners
use technology to make judgments, rather than being judged by technology.*

One technique we will definitely see more of in coming years is online streaming
of live coding sessions [Haar2017]. This has most of the benefits discussed in Section
8.1, and when combined with collaborative note-taking (Section 9.7) it can be a
close approximation to an in-class experience.

Looking even further ahead, [Ijss2000] identified four levels of online presence,
from realism (we can't tell the difference) through immersion (we forget the difference)
and involvement (we're engaged but aware of the difference) to suspension of
disbelief (we are doing most of the work). Crucially, they distinguish physical presence,
which is the sense of actually being somewhere, and social presence, which
is the sense of being with others. The latter is more important in most learning situations,
and again, we can foster it by using learners' everyday technology in the
classroom. For example, [Deb2018] found that real-time feedback on in-class exercises
using learners' own mobile devices improved concept retention and learner
engagement while reducing failure rates.

Online and asynchronous teaching are both still in their infancy. Centralized
MOOCs may prove to be an evolutionary dead end, but there are still many other
promising models to explore. In particular, [Broo2016] describes fifty ways that
groups can discuss things productively, only a handful of which are widely known
or implemented online. If we go where our learners are technologically rather than
requiring them to come to us, we may wind up learning as much as they do.

11.5 EXERCISES

TWO-WAY VIDEO (PAIRS/10)

Record a 2–3 minute video of yourself doing something, then swap machines with a
partner so that each of you can watch the other's video at 4x speed. How easy is it to
follow what's going on? What if anything did you miss?

VIEWPOINTS (INDIVIDUAL/10)

According to [Irib2009], different disciplines focus on different factors affecting the
success or otherwise of online communities:

Business: customer loyalty, brand management, extrinsic motivation.

Psychology: sense of community, intrinsic motivation.

Sociology: group identity, physical community, social capital, collective action.

Computer Science: technological implementation.

Which of these perspectives most closely corresponds to your own? Which are
you least aligned with?

HELPING OR HARMING (SMALL GROUPS/30)

Susan Dynarski's article in the *New York Times*[15] explains how and why schools are putting students who fail in-person courses into online courses, and how this sets them up for even further failure. Read the article and then:

1. In small groups, come up with 2–3 things that schools could do to compensate for these negative effects and create rough estimates of their per-learner costs.
2. Compare your suggestions and costs with those of other groups. How many full-time teaching positions do you think would have to be cut in order to free up resources to implement the most popular ideas for 100 learners?
3. As a class, do you think that would be a net benefit for the learners or not?

Budgeting exercises like this are a good way to tell who's serious about educational change. Everyone can think of things they'd like to do; far fewer are willing to talk about the tradeoffs needed to make change happen.

---

[15]https://www.nytimes.com/2018/01/19/business/online-courses-are-harming-the-students-who-need-the-most-help.html

12 Exercise Types

Every good carpenter has a set of screwdrivers, and every good teacher has different kinds of exercises to check what learners are actually learning, help them practice their new skills, and keep them engaged. This chapter starts by describing several kinds of exercises you can use to check if your teaching has been effective. It then looks at the state of the art in automated grading, and closes by exploring discussion, projects, and other important kinds of work that require more human attention to assess. Our discussion draws in part on the Canterbury Question Bank[1] [Sand2013], which has entries for various languages and topics in introductory computing.

12.1 THE CLASSICS

As Section 2.1 discussed, *multiple choice questions* (MCQs) are most effective when the wrong answers probe for specific misconceptions. They are usually designed to test the lower levels of Bloom's Taxonomy (Section 6.2), but can also require learners to exercise judgment.

> ### *A Multiple Choice Question*
> *In what order do operations occur when the computer evaluates the expression* `price = addTaxes(cost - discount)`*?*
> 1. *subtraction, function call, assignment*
> 2. *function call, subtraction, assignment*
> 3. *function call, then assignment and subtraction simultaneously*
> 4. *none of the above*

The second classic type of programming exercise is *code and run* (C&R), in which the learner writes code that produces a specified output. C&R exercises can be as simple or as complex as the teacher wants, but when used in class, they should be brief and have only one or two plausible correct answers. It's often enough to ask novices to calculate and print a single value or call a specific function: experienced teachers often forget how hard it can be to figure out which parameters go where. For more advanced learners, figuring out which function to call is more engaging and a better gauge of their understanding.

> ### *Code & Run*
> *The variable* `picture` *contains a full-color image read from a file. Using one function, create a black and white version of the image and assign it to a new variable called* `monochrome`*.*

Write and run exercises can be combined with MCQs. For example, this MCQ can only be answered by running the Unix `ls` command:

---

[1] http://web-cat.org/questionbank/

Combining MCQ with Code & Run
You are in the directory /home. *Which of the following files is* not *in that directory?*

1. autumn.csv
2. fall.csv
3. spring.csv
4. winter.csv

C&Rs help people practice the skills they most want to learn, but they can be hard to assess: there can be lots of unexpected ways to get the right answer, and people will be demoralized if an automatic grading system rejects their code because it doesn't match the teacher's. One way to reduce how often this occurs is to assess only their output, but that doesn't give them feedback on how they are programming. Another is to give them a small test suite they can run their code against before they submit it (at which point it is run against a more comprehensive set of tests). Doing this helps them figure out if they have completely misunderstood the intent of the exercise before they do anything that they think might cost them grades.

Instead of writing code that satisfies some specification, learners can be asked to write tests to determine whether a piece of code conforms to a spec. This is a useful skill in its own right, and doing it may give learners a bit more sympathy for how hard their teachers work.

Inverting Code & Run
The function monotonic_sum *calculates the sum of each section of a list of numbers in which the values are strictly increasing. For example, given the input* [1, 3, 3, 4, 5, 1]*, the output is* [4, 12, 1]*. Write and run unit tests to determine which of the following bugs the function contains:*

- *Considers every negative number the start of a new sub-sequence.*
- *Does not include the first value of each sub-sequence in the sub-sum.*
- *Does not include the last value of each sub-sequence in the sub-sum.*
- *Only re-starts the sum when values decrease rather than fail to increase.*

Fill in the blanks is a refinement of C&R in which the learner is given some starter code and has to complete it. (In practice, most C&R exercises are actually fill in the blanks because the teacher provides comments to remind the learners of the steps they should take.) Questions of this type are the basis for faded examples; as discussed in Chapter 4, novices often find them less intimidating than writing all the code from scratch, and since the teacher has provided most of the answer's structure, submissions are much more predictable and therefore easier to check.

Fill in the Blanks
Fill in the blanks so that the code below prints the string 'hat'.

```
text = 'all that it is'
slice = text[____:____]
print(slice)
```

Parsons Problems also avoid the "blank screen of terror" problem while allow-ing learners to concentrate on control flow separately from vocabulary [Pars2006; Eric2015; Morr2016; Eric2017]. Tools for building and doing Parsons Problems online exist [Ihan2011], but they can be emulated (albeit somewhat clumsily) by asking learners to rearrange lines of code in an editor.

Parsons Problem

Rearrange and indent these lines to sum the positive values in a list. (You will need to add colons in appropriate places as well.)

```
total = 0
if v > 0
total += v
for v in values
```

Note that giving learners more lines than they need, or asking them to rear-range some lines and add a few more, makes Parsons Problems significantly harder [Harm2016].

12.2 TRACING

Tracing execution is the inverse of a Parsons Problem: given a few lines of code, the learner has to trace the order in which those lines are executed. This is an essential debugging skill and a good way to solidify learners' understanding of loops, con-ditionals, and the evaluation order of function and method calls. The easiest way to implement it is to have learners write out a sequence of labeled steps. Having them choose the correct sequence from a set (i.e., presenting this as an MCQ) adds cogni-tive load without adding value, since they have to do all the work of figuring out the correct sequence, then search for it in the list of options.

Tracing Execution Order

In what order are the labeled lines in this block of code executed?

```
A)      vals = [-1, 0, 1]
B)      inverse_sum = 0
        try:
            for v in vals:
C)                inverse_sum += 1/v
        except:
D)          pass
```

Tracing values is similar to tracing execution, but instead of spelling out the order in which code is executed, the learner lists the values that one or more variables take on as the program runs. One way to implement this is to give the learner a table whose columns are labeled with variable names and whose rows are labeled with line numbers, and ask them to fill in the values taken on by the variables on those lines.

Tracing Values

What values do `left` *and* `right` *take on as this program executes?*

A) `left = 23`
B) `right = 6`
C) `while right:`
D)   `left, right = right, left % right`

Line	left	right

You can also require learners to trace code backwards to figure out what the input must have been for the code to produce a particular result [Armo2008]. These *reverse execution* problems require search and deductive reasoning, and when the output is an error message, they help learners develop valuable debugging skills.

Reverse Execution

Fill in the missing number in `values` *that caused this function to crash.*

```
values = [ [1.0, -0.5], [3.0, 1.5], [2.5, ___] ]
runningTotal = 0.0
for (reading, scaling) in values:
    runningTotal += reading / scaling
```

Minimal fix exercises also help learners develop debugging skills. Given a few lines of code that contain a bug, the learner must find it and make one small change to fix it. Making the change can be done using C&R, while identifying it can be done as a multiple choice question.

Minimal Fix

This function is supposed to test whether a number lies within a range. Make one small change so that it actually does so.

```
def inside(point, lower, higher):
    if (point <= lower):
        return false
    elif (point <= higher):
        return false
    else:
        return true
```

Theme and variation exercises are similar, but the learner is asked to make a small alteration that changes the output in some specific way instead of making a change to fix a bug. Allowed changes can include changing a variable's initial value, replacing one function call with another, swapping inner and outer loops, or changing the order of tests in a complex conditional. Again, this kind of exercise gives learners a chance to practice a useful real-world skill: the fastest way to produce the code they need is to to tweak code that already does something close.

Theme and Variations

Change the inner loop in the function below so that it fills the upper left triangle of an image with a specified color.

```
function fillTriangle(picture, color) is
    for x := 1 to picture.width do
        for y := 1 to picture.height do
            picture[x, y] = color
        end
    end
end
```

Refactoring exercises are the complement of theme and variation exercises: given a working piece of code, the learner has to modify it in some way *without* changing its output. For example, the learner could replace loops with vectorized expressions or simplify the condition in a while loop. This is also a useful real-world skill, but there are often so many ways to refactor code that grading requires human inspection.

Refactoring

Write a single list comprehension that has the same effect as this loop.

```
result = []
for v in values:
    if len(v) > threshold:
        result.append(v)
```

12.3 DIAGRAMS

Having learners draw concept maps and other diagrams gives insight into how they're thinking (Section 3.1), but free-form diagrams take human time and judgment to assess. *Labeling diagrams*, on the other hand, is almost as powerful pedagogically but much easier to scale.

Rather than having learners create diagrams from scratch, provide them with a diagram and a set of labels and have them put the latter in the right places on the former. The diagram can be a data structure ("after this code is executed, which variables point to which parts of this structure?"), a chart ("match each of these pieces of code with the part of the chart it generated"), or the code itself ("match each term to an example of that program element").

Labeling a Diagram

Figure 12.1 shows how a small fragment of HTML is represented in memory. Put the labels 1–9 on the elements of the tree to show the order in which they are reached in a depth-first traversal.

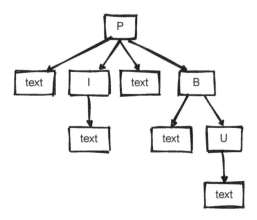

Figure 12.1: Labeling a diagram

Another way to use diagrams is to give learners the pieces of the diagram and ask them to arrange them correctly. This is a visual equivalent of a Parsons Problem, and you can provide as much or as little of a skeleton to help with placement as you think they're ready for. I have fond memories of trying to place resistors and capacitors in a circuit diagram in order to get the right voltage at a certain point, and have seen teachers give learners a fixed set of Scratch blocks and ask them to create a particular drawing using only those blocks.

Matching problems can be thought of as a special case of labeling in which the "diagram" is a column of text and the labels are taken from the other column. *One-to-one matching* gives the learner two lists of equal length and asks them to pair corresponding items, e.g., "match each piece of code with the output it produces."

Matching
Match each regular expression operator in Figure 12.2 with what it does.

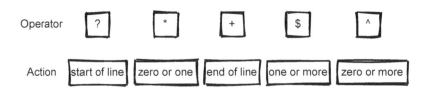

Figure 12.2: Matching items

With *many-to-many matching* the lists aren't the same length, so some items may be matched to several others and others may not be matched at all. Many-to-many are more difficult because learners can't do easy matches first to reduce their search space. Matching problems can be implemented by having learners submit lists of matching pairs (such as "A3, B1, C2"), but that's clumsy and error-prone. Having them recognize a set of correct pairs in an MCQ is even worse, as it's painfully easy to misread. Drawing or dragging works much better, but may require some work to implement.

Ranking is a special case of matching that is (slightly) more amenable to answering via lists, since our minds are pretty good at detecting errors or anomalies in sequences. The ranking criteria determine the level of reasoning required. If you have learners order sorting algorithms from fastest to slowest you are probably exercising recall (i.e., asking them recognizing the algorithms' names and know their properties), while asking them to rank solutions from most robust to most brittle exercises reasoning and judgment.

Summarization also requires learners to use higher-order thinking and gives them a chance to practice a skill that is very useful when reporting bugs. For example, you can ask learners, "Which sentence best describes how the output of f changes as x varies from 0 to 10?" and then given several options as a multiple choice question. You can also ask for very short free-form answers to questions in constrained domains, such as, "What is the key feature of a stable sorting algorithm?" We can't fully automate checks for these without a frustrating number of false positives (accepting wrong answers) and false negatives (rejecting correct ones), but questions of this kind lend themselves well to peer grading (Section 5.3).

12.4 AUTOMATIC GRADING

Automatic program grading tools have been around longer than I have been alive: the earliest published mention dates from 1960 [Holl1960], and the surveys published in [Douc2005; Ihan2010] mention many specific tools by name. Building such tools is a lot more complex than it might first seem. How are assignments represented? How are submissions tracked and reported? Can learners co-operate? How can submissions be executed safely? [Edwa2014a] is an entire paper devoted to an adaptive scheme for detecting and managing infinite loops in code submissions, and that's just one of the many issues that comes up.

When discussing auto-graders, it is important to distinguish learner satisfaction from learning outcomes. For example, [Magu2018] switched informal programming labs for a second-year CS course to a weekly machine-evaluated test using an auto-grader. Learners didn't like the automated system, but the overall failure rate for the course was halved and the number of learners gaining first class honors tripled. In contrast, [Rubi2014] also began to use an auto-grader designed for competitions, but saw no significant decrease in their learners' dropout rates; once again, learners made some negative comments about the tool, which the authors attribute to the quality of its feedback messages rather than to dislike of auto-grading.

[Srid2016] took a different approach. They used **fuzz testing** (i.e., randomly generated test cases) to check whether learner code does the same thing as a reference implementation supplied by the teacher. In the first project of a 1400-learner introductory course, fuzz testing caught errors that were missed by a suite of hand-written test cases for more than 48% of learners.

[Basu2015] gave learners a suite of solution test cases, but learners had to unlock each one by answering questions about its expected behavior before they were allowed to apply it to their proposed solution. For example, suppose learners had to write a function to find the largest adjacent pair of numbers in a list. Before being allowed to use the question's tests, they had to choose the right answer to, "What does largestPair(4, 3, -1, 5, 3, 3) produce?" In a 1300-person university course, the vast majority of learners chose to validate their understanding of test cases this way before attempting to solve problems, and then asked fewer questions and expressed less confusion about assignments.

Against Off-the-Shelf Tools

It's tempting to use off-the-shelf style checking tools to grade learners' code. However, [Nutb2016] initially found no correlation between human-provided marks and style-checker rule violations. Sometimes this was because learners violated one rule many times (thereby losing more points than they should have), but other times it was because they submitted the assignment starter code with few alterations and got more points than they should have.

Even tools built specifically for teaching can fall short of teachers' needs. [Keun2016a; Keun2016b] looked at the messages produced by 69 auto-grading tools. They found that the tools often do not give feedback on how to fix problems and take the next step. They also found that most teachers cannot easily adapt most of the tools to their needs: like many workflow tools, they tend to enforce their creators' unrecognized assumptions about how institutions work. Their classification scheme is a useful shopping list when looking at tools of this kind.

[Buff2015] presents a well-informed reflection on the whole idea of providing automated feedback. Their starting point is that, "Automated grading systems help learners identify bugs in their code, [but] may inadvertently discourage learners from thinking critically and testing thoroughly and instead encourage dependence on the teacher's tests." One of the key issues they identified is that a learner may thoroughly test their code, but the feature may still not be implemented according to the teacher's specifications. In this case, the "failure" is not caused by a lack of testing but by a misunderstanding of the requirements, and it is unlikely that more testing will expose the problem. If the auto-grading system doesn't provide insightful, actionable feedback, this experience will only frustrate the learner.

In order to provide that feedback, [Buff2015]'s system identifies which methods in the learner's code are executed by failing tests so that the system can associate failed tests with particular features within the learner's submission. The system decides whether specific hints have been "earned" by seeing whether the learner has tested the associated feature enough, so learners cannot rely on hints instead of doing tests.

[Srid2016] describes some other approaches for sharing feedback with learners when automatically testing their code. The first is to provide the expected output for the tests—but then learners hard-code output for those inputs (because anything that can be gamed will be). The second is to report the pass/fail results for the learners' code, but only supply the actual inputs and outputs of the tests after the submission date. However, telling learners that they are wrong but not telling them why is frustrating.

A third option is to use a technique called **hashing** to generate a value that depends on the output but doesn't reveal it. If the user produces exactly the right output then its hash will unlock the solution, but it is impossible to work backward from the hash to figure out what the output is supposed to be. Hashing requires more work and explanation to set up, but strikes a good balance between revealing answers prematurely and not revealing them when it would help.

12.5 HIGHER-LEVEL THINKING

Many other kinds of programming exercises are hard for teachers to assess in a class with more than handful of learners and equally hard for automated platforms to assess at all. Larger programming projects are (hopefully) what classes are building toward, but the only way to give feedback is case by case.

Code review is also hard to grade automatically in general, but can be tackled if learners are given a list of faults to look for and asked to match particular comments against particular lines of code. For example, the learner can be told that there are two indentation errors and one bad variable name and asked to point them out. If they are more advanced, they could be given half a dozen kinds of remarks they could make about the code without being told how many of each they should find.

[Steg2016b] is a good starting point for a code style rubric, while [Luxt2009] looks at peer review in programming classes more generally. If you are going to have learners do reviews, use calibrated peer review (Section 5.3) so that they have models of what good feedback should look like.

Code Review
Mark the problems in each line of code using the rubric provided.

```
01)   def addem(f):
02)       x1 = open(f).readlines()
03)       x2 = [x for x in x1 if x.strip()]
04)       changes = 0
05)       for v in x2:
06)           print('total', total)
07)               tot = tot + int(v)
08)       print('total')
```

1. *poor variable name*   2. *use of undefined variable*
3. *missing return value*   4. *unused variable*

12.6   EXERCISES

CODE AND RUN (PAIRS/10)

Create a short C&R exercise, then trade with a partner and see how long it takes each of you to understand and do the other's exercise. Were there any ambiguities or misunderstandings in the exercise description?

INVERTING CODE AND RUN (SMALL GROUPS/15)

Form groups of 4–6 people. Have each member of the group create an inverted C&R exercise that requires people to figure out what input produces a particular output. Pick two at random and see how many different inputs the group can find that satisfy the requirements.

TRACING VALUES (PAIRS/10)

Write a short program (10–15 lines), trade with a partner, and trace how the variables in the program change value over time. What differences are there in how you and your partner wrote down your traces?

REFACTORING (SMALL GROUPS/15)

Form groups of 3–4 people. Have each person select a short piece of code (10–30 lines long) they have written that isn't as tidy as it could be, then choose one at random and have everyone in the group tidy it up independently. How do your cleaned-up versions differ? How well or how poorly would you be able to accommodate all of these variations if marking automatically or in a large class?

LABELING A DIAGRAM (PAIRS/10)

Draw a diagram showing something that you have explained recently: how browsers fetch data from servers, the relationship between objects and classes, or how data frames are indexed in R. Put the labels on the side and ask your partner to place them.

PENCIL-AND-PAPER PUZZLES (WHOLE CLASS/15)

[Butl2017] describes a set of pencil-and-paper puzzles that can be turned into introductory programming assignments and reports that these assignments are enjoyed by learners and encourage meta-cognition. Think of a simple pencil-and-paper puzzle or game you played as a child and describe how you would turn it into a programming exercise.

COUNTING FAILURES (PAIRS/15)

Any useful estimate of how much time an exercise needs must take into account how frequent failures are and how much time is lost to them. For example, editing text files seems like a simple task, but what about finding those files? Most GUI editors save things to the user's desktop or home directory; if the files used in a course are stored somewhere else, a substantial fraction won't be able to navigate to the right directory without help. (If this seems like a small problem to you, please revisit the discussion of expert blind spot in Chapter 3.)

Working with a partner, make a list of "simple" things you have seen go wrong in exercises you have used or taken. How often do they come up? How long do they take learners to fix on their own or with help? How much time do you currently budget in class to deal with them?

SPEAKING OF TIMINGS (INDIVIDUAL/10)

How accurate have the time estimates on the exercises in this book been so far?

13 Building a Community of Practice

You don't have to fix all of society's ills in order to teach programming, but you *do* have to be involved in what happens outside your class if you want people to learn. This applies to teachers as well as learners: many free-range teachers start as volunteers or part-timers and have to juggle many other commitments. What happens outside the classroom is as important to their success as it is to their learners', so the best way to help both is to foster a teaching community.

Finland and Why Not

Finland's schools are among the most successful in the world, but as Anu Partanen pointed out[1], they haven't succeeded in isolation. Other countries' attempts to adopt Finnish teaching methods are doomed to fail unless those countries also ensure that children (and their parents) are safe, well nourished, and treated fairly by the courts [Sahl2015; Wilk2011]. This isn't surprising given what we know about the importance of motivation for learning (Chapter 10): everyone does worse if they believe the system is unpredictable, unfair, or indifferent.

A framework for thinking about teaching communities is **situated learning**, which focuses on how **legitimate peripheral participation** leads to people becoming members of a **community of practice** [Weng2015]. Unpacking those terms, a community of practice is a group of people bound together by interest in some activity, such as knitting or particle physics. Legitimate peripheral participation means doing simple, low-risk tasks that the community recognizes as valid contributions: making your first scarf, stuffing envelopes during an election campaign, or proofreading documentation for open source software.

Situated learning focuses on the transition from being a newcomer to being accepted as a peer by those who are already community members. This typically means starting with simple tasks and tools, then doing similar tasks with more complex tools, and finally tackling the same work as advanced practitioners. For example, children learning music may start by playing nursery rhymes on a recorder or ukulele, then play other simple songs on a trumpet or saxophone in a band, and finally start exploring their own musical tastes. Common ways to support this progression include:

---

[1] https://www.theatlantic.com/national/archive/2011/12/what-americans-keep-ignoring-about-finlands-school-success/250564/

Problem solving: "I'm stuck—can we work on designing this lesson together?"

Requests for information: "What's the password for the mailing list manager?"

Seeking experience: "Has anyone had a learner with a reading disability?"

Sharing assets: "I put together a website for a class last year that you can use as a starting point."

Coordination: "Can we combine our t-shirt orders to get a discount?"

Building an argument: "It will be easier to convince my boss to make changes if I know how other bootcamps do this."

Documenting projects: "We've had this problem five times now. Let us write it down once and for all."

Mapping knowledge: "What other groups are doing things like this in nearby neighborhoods or cities?"

Visits: "Can we come and see your after-school program? We need to establish one in our city."

Broadly speaking[2], a community of practice can be a:

Community of action: people focused on a shared goal, such as getting someone elected.

Community of concern: members are brought together by a shared issue, such as dealing with a long-term illness.

Community of interest: focused on a shared love of something like backgammon or knitting.

Community of place: people who happen to live or work side by side.

Most communities are mixes of these, such as people in Toronto who like teaching tech. A community's focus can also change over time: for example, a support group for people dealing with depression (community of concern) can decide to raise funds to keep a help line going (community of action). Running the help line can then become the group's focus (community of interest).

Soup, Then Hymns

Manifestos are fun to write, but most people join a volunteer community to help and be helped rather than to argue over the wording of a vision statement[3]. You should therefore focus on what people can create that will be used by other community members right away. Once your organization shows that it can achieve small things, people will be more confident that it's worth helping you with bigger projects. That's the time to worry about defining the values that will guide your members.

---

[2]https://www.feverbee.com/types-of-community-and-activity-within-the-community/

[3]People who prefer the latter are often *only* interested in arguing...

13.1 LEARN, THEN DO

The first step in building a community is to decide if you should, or whether you would be more effective joining an existing organization. Thousands of groups are already teaching people tech skills, from the 4-H Club[4] and literacy programs[5] to get-into-coding nonprofits like Black Girls Code[6] and Bridge[7]. Joining an existing group will give you a head start on teaching, an immediate set of colleagues, and a chance to learn more about how to run things; hopefully, learning those skills while making an immediate contribution will be more important than being able to say that you're the founder or leader of something new.

Whether you join an existing group or set up one of your own, you will be more effective if you do a bit of background reading on community organizing. [Alin1989; Lake2018] is probably the best-known work on grassroots organizing, while [Brow2007; Midw2010; Lake2018] are practical manuals rooted in decades of practice. If you want to read more deeply, [Adam1975] is a history of the Highlander Folk School, whose approach has been emulated by many successful groups, while [Spal2014] is a guide to teaching adults written by someone with deep personal roots in organizing and NonprofitReady.org[8] offers free professional development training.

13.2 FOUR STEPS

Everyone who gets involved with your organization (including you) goes through four phases: recruitment, onboarding, retention, and retirement. You don't need to worry about this cycle when you're getting started, but it is worth thinking about as soon as more than a handful of non-founders are involved.

The first step is recruiting volunteers. Your marketing should help you with this by making your organization findable and by making its mission and value clear to people who might want to get involved (Chapter 14). Share stories that exemplify the kind of help you want as well as stories about the people you're helping, and make it clear that there are many ways to get involved. (We discuss this in more detail in the next section.)

Your best source of new recruits is your own classes: "see one, do one, teach one" has worked well for volunteer organizations for as long as there have *been* volunteer organizations. Make sure that every class or other encounter ends by telling people how they could help and that their help would be welcome. People who come to you this way will know what you do and have recent experience of being on the receiving end of what you offer, which helps your organization avoid collective expert blind spot (Chapter 3).

---

[4]http://www.4-h-canada.ca/

[5]https://www.frontiercollege.ca/

[6]http://www.blackgirlscode.com/

[7]http://bridgeschool.io/

[8]https://www.nonprofitready.org/

Start Small

Ben Franklin[9] observed that a person who has performed a favor for someone is more likely to do them another favor than someone who had received a favor from that person. Asking people to do something small for you is therefore a good step toward getting them to do something larger. One natural way to do this when teaching is to ask people to submit fixes for your lesson materials for typos or unclear wording, or to suggest new exercises or examples. If your materials are written in a maintainable way (Section 6.3), this gives them a chance to practice some useful skills and gives you an opportunity to start a conversation that might lead to a new recruit.

The middle of the volunteer lifecycle is onboarding and retention, which we will cover in Sections 13.3 and 13.4. The final step is retirement: everyone moves on eventually, and healthy organizations plan for this. A few simple things can make both the person leaving and everyone who is staying feel positive about the change:

Ask people to be explicit about their departure so that everyone knows they've actually left.

Make sure they don't feel embarrassed or ashamed about leaving or about anything else.

Give them an opportunity to pass on their knowledge. For example, you can ask them to mentor someone for a few weeks as their last contribution, or to be interviewed by someone who is staying with the organization to collect any stories that are worth re-telling.

Make sure they hand over the keys. It's awkward to discover six months after someone has left that they're the only person who knows how to book a field for the annual picnic.

Follow up 2–3 months after they leave to see if they have any further thoughts about what worked and what didn't while they were with you, or any advice to offer that they either didn't think to give or were uncomfortable giving on their way out the door.

Thank them, both when they leave and the next time your group gets together.

A Missing Manual

Thousands of books have been written on how to start a company. Only a handful describe how to end one or leave one gracefully, even though there is an ending for every beginning. If you ever write one, please let me know.

13.3 ONBOARDING

After deciding to become part of a group, people need to get up to speed, and [Shol2019] summarizes what we know about doing this. The first rule is to have

---

[9]https://en.wikipedia.org/wiki/Ben_Franklin_effect

and enforce a Code of Conduct (Section 9.1), and to find an independent party who is willing to receive and review reports of inappropriate behavior. Someone outside the organization will have the objectivity that organization members may lack, and can protect reporters who might hesitate to raise issues concerning project leaders out of fear of retribution or damage to their reputation. Project leaders should also publicize the enforcement decisions so that the community recognizes that the code is meaningful.

The next most important thing is to be welcoming. As Fogel said [Foge2005], "If a project doesn't make a good first impression, newcomers may wait a long time before giving it a second chance." Other authors have empirically confirmed the importance of kind and polite social environments in open projects [Sing2012; Stei2013; Stei2018]:

Post a welcome message on the project's social media pages, Slack channels, forums, or email lists. Projects might consider maintaining a dedicated "Welcome" channel or list, where a project lead or community manager writes a short post asking newcomers to introduce themselves.

Help people find ways to make an initial contribution, such as labeling particular lessons or workshops that need work as "suitable for newcomers" and asking established members not to fix them in order to ensure there are suitable places for new arrivals to start work.

Direct the newcomer to project members like them to demonstrate that they belong.

Point the newcomer to essential project resources such as the contribution guidelines.

Designate one or two members to serve as a point of contact for each newcomer. Doing this can make the newcomer less reluctant to ask questions.

A third rule that helps everyone (not just newcomers) is to make knowledge findable and keep it up to date. Newcomers are like explorers who must orient themselves within an unfamiliar landscape [Dage2010]. Information that is spread out usually makes newcomers feel lost and disoriented. Given the different possibilities of places to maintain information (e.g., wikis, files in version control, shared documents, old tweets or Slack messages, and email archives) it is important to keep information about a specific topic consolidated in a single place so that newcomers do not need to navigate multiple data sources to find what they need. Organizing the information make newcomers more confident and oriented [Stei2016].

Finally, acknowledge newcomers' first contributions and figure out where and how they might help in the longer term. Once they have carried their first contribution over the line, you and they are likely to have a better sense of what they have to offer and how the project can help them. Help newcomers find the next problem they might want to work on or point them at the next thing they might enjoy reading. In particular, encouraging them to help the next wave of newcomers is both a good way to recognize what they have learned and an effective way to pass it on.

13.4 RETENTION

If your people aren't having a ball doing it, there is something very wrong.
— Saul Alinsky

Community members shouldn't expect to enjoy every moment of their work with your organization, but if they don't enjoy any of it, they won't stick around. Enjoyment doesn't necessarily mean having an annual party: people may enjoy cooking, coaching, or just working quietly beside others. There are several things every organization should do to ensure that people are getting something they value out of their work:

Ask people what they want rather than guessing. Just as you are not your learners (Section 6.1), you are probably different from other members of your organization. Ask people what they want to do, what they're comfortable doing (which may not be the same thing), and what constraints there are on their time. They might say, "Anything," but even a short conversation will probably uncover the fact that they like interacting with people but would rather not be managing the group's finances or vice versa.

Provide many ways to contribute. The more ways there are for people to help, the more people will be able to. Someone who doesn't like standing in front of an audience may be able to maintain your organization's website, handle its accounts, or proofread lessons.

Recognize contributions. Everyone likes to be appreciated, so communities should acknowledge their members' contributions both publicly and privately by mentioning them in presentations, putting them on the website, and so on. Every hour that someone has given your project may be an hour taken away from their personal life or their official employment; recognize that fact and make it clear that while more hours would be welcome, you do not expect them to make unsustainable sacrifices.

Make space. You think you're being helpful, but intervening in every decision robs people of their autonomy, which in return reduces their motivation (Section 10). In particular, if you're always the first one to reply to email or chat messages, people have less opportunity to grow as members and to create horizontal collaborations. As a result, the community will continue to be centered around one or two individuals rather than becoming a highly connected network in which others feel comfortable participating.

Another way to reward participation is to offer training. Organizations need budgets, grant proposals, and dispute resolution. Most people are never taught how to do this any more than they are taught how to teach, so the opportunity to gain transferable skills is a powerful reason for people to get and stay involved. If you are going to do this, don't try to provide the training yourself unless it's what you specialize in. Many civic and community groups have programs of this kind, and you can probably make a deal with one of them.

Finally, while volunteers can do a lot, tasks like system administration and accounting eventually need paid staff. When you reach this point, either pay people nothing or pay them a proper wage. If you pay them nothing, their real reward is the satisfaction of doing good; if you pay them a token amount, on the other hand, you take that away without giving them the satisfaction of earning a living.

13.5 GOVERNANCE

Every organization has a power structure: the only question is whether it's formal and accountable or informal and therefore unaccountable [Free1972]. The latter actually works pretty well for groups of up to half a dozen people in which everyone knows everyone else. Beyond that, you need rules to spell out who has the authority to make which decisions and how to achieve consensus (Section E.1).

The governance model I prefer is a **commons**, which is something managed jointly by a community according to rules they themselves have evolved and adopted [Ostr2015]. As [Boll2014] emphasizes, all three parts of that definition are essential: a commons isn't just a shared pasture or a set of software libraries, but also includes the community that shares it and the rules they use to do so.

For-profit corporations and incorporated nonprofits are more popular models; the mechanics vary from jurisdiction to jurisdiction, so you should seek advice before choosing[10]. Both kinds of organization vest ultimate authority in their board. Broadly speaking, this is either a **service board** whose members also take on other roles in the organization or a **governance board** whose primary responsibility is to hire, monitor, and if need be fire the director. Board members can be elected by the community or appointed; in either case, it's important to prioritize competence over passion (the latter being more important for the rank and file) and to try to recruit for particular skills such as accounting, marketing, and so on.

Choose Democracy

When the time comes, make your organization a democracy: sooner or later (usually sooner), every appointed board turns into a mutual agreement society. Giving your members power is messy, but is the only way invented so far to ensure that organizations continue to meet people's actual needs.

13.6 LOOK AFTER YOURSELF

Burnout is a chronic risk in any community activity [Pign2016], so learn to say no more often than you say yes. If you don't take care of yourself, you won't be able to take care of your community.

One way to make your "no" stick is to write a to-don't list of things that would be worth doing but which you *aren't* going to do. At the time of writing, mine includes

---

[10]This is one of the times when having ties with local government or other like-minded organizations pays off.

four books, two software projects, redesign of my personal website, and learning to play the penny whistle.

Finally, remind yourself every now and then that every organization eventually needs fresh ideas and leadership. When that time comes, train your successors and move on as gracefully as you can. They will undoubtedly do things you wouldn't have, but few things in life are as satisfying as watching something you helped build take on a life of its own. Celebrate that—you won't have any trouble finding something else to keep you busy.

13.7 EXERCISES

Several of these exercises are taken from [Brow2007].

WHAT KIND OF COMMUNITY? (INDIVIDUAL/15)

Re-read the description of the four types of communities and decide which one(s) your group is or aspires to be.

PEOPLE YOU MAY MEET (SMALL GROUPS/30)

As an organizer, part of your job is sometimes to help people find a way to contribute despite themselves. In small groups, pick three of the people below and discuss how you would help them become a better contributor to your organization.

Anna knows more about every subject than everyone else put together—at least, she thinks she does. No matter what you say, she'll correct you; no matter what you know, she knows better.

Catherine has so little confidence in her own ability that she won't make any decision, no matter how small, until she has checked with someone else.

Frank enjoys knowing things that other people don't. He can work miracles, but when asked how he did it, he'll grin and say, "Oh, I'm sure you can figure it out."

Hediyeh is quiet. She never speaks up in meetings, even when she knows other people are wrong. She might contribute to the mailing list, but she's very sensitive to criticism and always backs down instead of defending her point.

Ken takes advantage of the fact that most people would rather shoulder his share of the work than complain about him. The frustrating thing is that he's so plausible when someone finally does confront him. "There have been mistakes on all sides," he says, or, "Well, I think you're nit-picking."

Melissa means well, but somehow something always comes up, and her tasks are never finished until the last possible moment. Of course, that means that everyone who is depending on her can't do their work until *after* the last possible moment...

Raj is rude. "It's just the way I talk," he says. "If you can't hack it, go find another team." His favorite phrase is, "That's stupid," and he uses an obscenity in every second sentence.

VALUES (SMALL GROUPS/45)

Answer these questions on your own, then compare your answers with others'.

1. What are the values your organization expresses?
2. Are these the values you want the organization to express?
3. If not, what values would you like it to express?
4. What are specific behaviors that demonstrate those values?
5. What behaviors would demonstrate the opposite of those values?

MEETING PROCEDURES (SMALL GROUPS/30)

Answer these questions on your own, then compare your answers with others'.

1. How are your meetings run?
2. Is this how you want your meetings to be run?
3. Are the rules for running meetings explicit or just assumed?
4. Are these the rules you want?
5. Who is eligible to vote or make decisions?
6. Is this who you want to be vested with decision-making authority?
7. Do you use majority rule, make decisions by consensus, or something else?
8. Is this the way you want to make decisions?
9. How do people in a meeting know when a decision has been made?
10. How do people who weren't at a meeting know what decisions were made?
11. Is this working for your group?

SIZE (SMALL GROUPS/20)

Answer these questions on your own, then compare your answers with others'.

1. How big is your group?
2. Is this the size you want for your organization?
3. If not, what size would you like it to be?
4. Do you have any limits on the size of membership?
5. Would you benefit from setting such a limit?

BECOMING A MEMBER (SMALL GROUPS/45)

Answer these questions on your own, then compare your answers with others'.

1. How does someone join your group?
2. How well does this process work?
3. Are there membership dues?
4. Are people required to agree to any rules of behavior upon joining?
5. Are these the rules for behavior you want?
6. How does a newcomer find out what needs to be done?
7. How well does this process work?

STAFFING (SMALL GROUPS/30)

Answer these questions on your own, then compare your answers with others'.

1. Do you have paid staff in your organization or is everyone a volunteer?
2. Should you have paid staff?
3. Do you want/need more or less staff?
4. What do the staff members do?
5. Are these the primary roles and functions that you need staff to fill?
6. Who supervises your staff?
7. Is this the supervision process that you want for your group?
8. What is your staff paid?
9. Is this the right salary to get the needed work done?

MONEY (SMALL GROUPS/30)

Answer these questions on your own, then compare your answers with others'.

1. Who pays for what?
2. Is this who you want to be paying?
3. Where do you get your money?
4. Is this how you want to get your money?
5. If not, do you have any plans to get it another way?
6. If so, what are they?
7. Who is following up to make sure that happens?
8. How much money do you have?
9. How much do you need?
10. What do you spend most of your money on?
11. Is this how you want to spend your money?

BORROWING IDEAS (WHOLE CLASS/15)

Many of my ideas about how to build a community have been shaped by my experience in open source software development. [Foge2005] (which is available online[11]) is a good guide to what has and hasn't worked for those communities, and the Open Source Guides site[12] has a wealth of useful information as well. Choose one section of the latter, such as "Finding Users for Your Project" or "Leadership and Governance," and give a two-minute presentation to the group of one idea from it that you found useful or that you strongly disagreed with.

---

[11] http://producingoss.com/

[12] https://opensource.guide/

WHO ARE YOU? (SMALL GROUPS/20)

The National Oceanic and Atmospheric Administration (NOAA) has a short, useful, and amusing guide to dealing with disruptive behaviors[13]. It categorizes those behaviors under labels like "talkative," "indecisive," and "shy," and outlines strategies for handling each. In groups of 3–6, read the guide and decide which of these descriptions best fits you. Do you think the strategies described for handling people like you are effective? Are other strategies equally or more effective?

CREATING LESSONS TOGETHER (SMALL GROUPS/30)

One of the keys to the success of the Carpentries[14] is their emphasis on building and maintaining lessons collaboratively [Wils2016; Deve2018]. Working in groups of 3–4:

1. Pick a short lesson you have all used.
2. Do a careful review to create a unified list of suggestions for improvements.
3. Offer those suggestions to the lesson's author.

ARE YOU CRISPY? (INDIVIDUAL/10)

Johnathan Nightingale wrote[15]:

> When I worked at Mozilla, we used the term "crispy" to refer to the state right before burnout. People who are crispy aren't fun to be around. They are curt. They are itching for a fight they can win. They cry without much warning. ... we would recognize crispiness in our colleagues and take care of each other [but] it is an ugly thing that we saw it so much that we had a whole cultural process around it.

Answer "yes" or "no" to each of the following questions. How close are you to burning out?

- Have you become cynical or critical at work?
- Do you have to drag yourself to work or do you have trouble getting started?
- Have you become irritable or impatient with co-workers?
- Do you find it hard to concentrate?
- Do you fail to get satisfaction from your achievements?
- Are you using food, drugs or alcohol to feel better or to simply not feel?

---

[13] https://coast.noaa.gov/ddb/story_html5.html

[14] http://carpentries.org

[15] https://mailchi.mp/d54702d0a790/take-my-horse-to-the-sand-hill-road

14 Outreach

It's fashionable in tech circles to disparage universities and government institutions as slow-moving dinosaurs, but in my experience they are no worse than companies of similar size. Your local school board, library, and your city councilor's office may be able to offer space, funding, publicity, connections with other groups that you may not have met yet, and a host of other useful things; getting to know them can help you solve or avoid problems in the short term and pay dividends down the road.

14.1 MARKETING

People with academic or technical backgrounds often think that **marketing** is about spin and misdirection. In reality, it's about seeing things from other people's perspective, understanding their wants and needs, and explaining how you can help them—in short, how to teach them. This chapter will look at how to use ideas from the previous chapters to get people to understand and support what you're doing.

The first step is to figure out what you are offering to whom, i.e., what actually brings in the volunteers, funding, and other support you need to keep going. The answer is often counter-intuitive. For example, most scientists think their papers are their product, but it's actually their grant proposals, because those are what brings in grant money [Kuch2011]. Their papers are the advertising that persuades people to fund those proposals, just as albums are now what persuades people to buy musicians' concert tickets and t-shirts.

Suppose that your group offers weekend programming workshops to people who are re-entering the workforce after being away for several years. If workshop participants can pay enough to cover your costs, then they are your customers and the workshops are the product. If, on the other hand, the workshops are free or the learners are only paying a token amount to cut the no-show rate, then your actual product may be some mix of:

- your grant proposals;
- the alumni of your workshops that the companies sponsoring you would like to hire;
- the half-page summary of your workshops in the mayor's annual report to city council that shows how she's supporting the local tech sector; or
- the personal satisfaction that volunteers get from teaching.

As with lesson design (Chapter 6), the first steps in marketing are to create personas of people who might be interested in what you're doing and to figure out which of their needs you can meet. One way to summarize the latter is to write **elevator pitches** aimed at different personas. A widely used template for these is:

For	*target audience*
who	*dissatisfaction with what's currently available*
our	*category*
provide	*key benefit.*
Unlike	*alternatives*
our program	*key distinguishing feature.*

Continuing the weekend workshop example, we could use this pitch for participants:

> For *people re-entering the workforce after being away for several years* who *still have family responsibilities*, our *introductory programming workshops* provide *weekend classes with on-site childcare*. Unlike *online classes*, our program *gives people a chance to meet others at the same stage of life*.

and this one for decision makers at companies that might sponsor the workshops:

> For *companies that want to recruit entry-level software developers* that *struggle to find mature candidates from diverse backgrounds* our *introductory programming workshops* provide *potential recruits*. Unlike *college recruiting fairs*, our program *connects companies with a wide variety of candidates*.

If you don't know why different potential stakeholders might be interested in what you're doing, ask them. If you do know, ask them anyway: answers can change over time, and you may discover things you previously overlooked.

Once you have these pitches, they should drive what you put on your website and in publicity material to help people figure out as quickly as possible if you and they have something to talk about. (You probably *shouldn't* copy them verbatim, though: many people in tech have seen this template so often that their eyes will glaze over if they encounter it again.)

As you are writing these pitches, remember that there are many reasons to learn how to program (Section 1.4). A sense of accomplishment, control over their own lives, and being part of a community may motivate people more than money (Chapter 10). They might volunteer to teach with you because their friends are doing it; similarly, a company may say that they're sponsoring classes for economically disadvantaged high school students because they want a larger pool of potential employees further down the road, but the CEO might actually be doing it simply because it's the right thing to do.

14.2 BRANDING AND POSITIONING

A **brand** is someone's first reaction to a mention of a product; if the reaction is "what's that?", you don't have a brand (yet). Branding is important because people aren't going to help something they don't know about or don't care about.

Most discussion of branding today focuses on how to build awareness online. Mailing lists, blogs, and Twitter all give you ways to reach people, but as the volume of misinformation increases, people pay less attention to each individual interruption. This makes **positioning** ever more important. Sometimes called "differentiation," it is what sets your offering apart from others—the "unlike" section of your elevator pitches. When you reach out to people who are already familiar with your field, you should emphasize your positioning, since it's what will catch their attention.

There are other things you can do to help build your brand. One is to use props like a robot that one of your learners made from scraps she found around the house [Schw2013] or the website another learner made for his parents' retirement home. Another is to make a short video—no more than a few minutes long—that showcases the backgrounds and accomplishments of your learners. The aim of both is to tell a story: while people always ask for data, they believe and remember stories.

Foundational Myths

One of the most compelling stories a person or group can tell is why and how they got started. Are you teaching what you wish someone had taught you but didn't? Was there one particular person you wanted to help, and that opened the floodgates? If there isn't a section on your website starting, "Once upon a time," think about adding one.

One crucial step is to make your organization findable in online searches. [DiSa2014b] discovered that the search terms that parents used for out-of-school computing classes didn't actually find those classes, and many other groups face similar challenges. There's a lot of folklore about how to make things findable (otherwise known as **search engine optimization** or SEO); given Google's near-monopoly powers and lack of transparency, most of it boils down to trying to stay one step ahead of algorithms designed to prevent people from gaming rankings.

Unless you're very well funded, the best you can do is to search for yourself and your organization on a regular basis and see what comes up, then read these guidelines[1] and do what you can to improve your site. Keep this XKCD cartoon[2] in mind: people don't want to know about your org chart or get a virtual tour of your site—they want your address, parking information, and some idea of what you teach, when you teach it, and how it's going to change their lives.

Not Everyone Lives Online

These examples assume people have access to the Internet and that groups have money, materials, free time, and/or technical skills. Many don't—in fact, those serving economically disadvantaged groups almost certainly don't. (As Rosario Robinson says, "Free works for those that can afford free.") Stories are more important than course outlines in those situations because they are easier to retell. Similarly, if the people you hope to reach are not online as often as you, then notice boards in schools, local libraries, drop-in centers, and grocery stores may be the most effective way to reach them.

---

[1] https://moz.com/learn/seo/on-page-factors

[2] https://xkcd.com/773/

14.3   THE ART OF THE COLD CALL

Building a website and hoping that people find it is easy; calling people up or knocking on their door without any sort of prior introduction is much harder. As with standing up and teaching, though, it's a craft that can be learned. Here are ten simple rules for talking people into things:

1: Don't. If you have to talk someone into something, odds are that they don't really want to do it. Respect that: it's almost always better in the long run to leave some particular thing undone than to use guilt or any underhanded psychological tricks that will only engender resentment.

2: Be kind. I don't know if there actually is a book called "Secret Tricks of the Ninja Sales Masters", but if there is, it probably tells readers that doing something for a potential customer that creates a sense of obligation, which in turn increases the odds of a sale. That may work, but it only works once and it's a skeezy thing to do. On the other hand, if you are genuinely kind and help other people because it's what good people do, you just might inspire them to be good people too.

3: Appeal to the greater good. If you open by talking about what's in it for them, you are signaling that they should think of their interaction with you as a commercial exchange of value to be bargained over. Instead, start by explaining how whatever you want them to help with is going to make the world a better place, and *mean it*. If what you're proposing isn't going to make the world a better place, propose something better.

4: Start small. Most people are understandably reluctant to dive into things head-first, so give them a chance to test the waters and to get to know you and everyone else involved in whatever it is you want help with. Don't be surprised or disappointed if that's where things end: everyone is busy or tired or has projects of their own, or maybe just has a different mental model of how collaboration is supposed to work. Remember the 90-9-1 rule—90% of people will watch, 9% will speak up, and 1% will actually do things—and set your expectations accordingly.

5: Don't build a project: build a community. I used to belong to a baseball team that never actually played baseball: our "games" were just an excuse for us to hang out and enjoy each other's company. You probably don't want to go quite that far, but sharing a cup of tea with someone or celebrating the birth of their first grandchild can get you things that no reasonable amount of money can.

6: Establish a point of connection. "I was speaking to X" or "we met at Y" gives them context, which in turn makes them more comfortable. This must be specific: spammers and cold-callers have trained us all to ignore anything that starts, "I recently came across your website..."

7: Be specific about what you are asking for. People need to know this so that they can figure out whether the time and skills they have are a match for what you need. Being realistic up front is also a sign of respect: if you tell people you need a hand moving a few boxes when you're actually packing up an entire house, they're probably not going to help you a second time.

8: Establish your credibility. Mention your backers, your size, how long your group has been around, or something that you've accomplished in the past so that they'll believe you're worth taking seriously.

9: Create a slight sense of urgency. "We're hoping to launch this in the spring" is more likely to get a positive response than "We'd eventually like to launch this." However, the word "slight" is important: if your request is urgent, most people will assume you're disorganized or that something has gone wrong and may then err on the side of prudence.

10: Take a hint. If the first person you ask for help says no, ask someone else. If the fifth or the tenth person says no, ask yourself if what you're trying to do makes sense and is worth doing.

The email template follows all of these rules. It has worked pretty well: we found that about half of emails were answered, about half of those wanted to talk more, and about half of those led to workshops, which means that 10–15% of targeted emails turned into workshops. That can still be pretty demoralizing if you're not used to it, but is much better than the 2–3% response rate most organizations expect with cold calls.

> Hi NAME,
> I hope you don't mind mail out of the blue, but I wanted to follow up on our conversation at VENUE to see if you would be interested having us run a teacher training workshop—we're scheduling the next batch over the next couple of weeks.
> This one-day workshop will teach your volunteers a handful of practical, evidence-based teaching practices. It has been run over a hundred times in various forms on six continents for nonprofit organizations, libraries, and companies, and all of the material is freely available online at http://teachtogether.tech. Topics will include:
> * learner personas
> * differences between different kinds of learners
> * using formative assessment to diagnose misunderstandings
> * teaching as a performance art
> * what motivates and demotivates adult learners
> * the importance of inclusivity and how to be a good ally
>
> If this sounds interesting, please give me a shout—I'd welcome a chance to talk ways and means.
> Thanks,
> NAME

Referrals

Building alliances with other groups that are doing things related to your work pays off in many ways. One of those is referrals: if someone who approaches you for help would be better served by some other organization, take a moment to make an introduction. If you've done this several times, add something to your website to help the next person find what they need. The organizations you are helping will soon start to help you in return.

14.4 ACADEMIC CHANGE

Everyone is afraid of the unknown and of embarrassing themselves. As a result, most people would rather fail than change. For example, Lauren Herckis looked at why university faculty don't adopt better teaching methods[3]. She found that the main reason is a fear of looking stupid in front of learners; secondary reasons were concern that the inevitable bumps in changing teaching methods would affect course evaluations (which could in turn affect promotion and tenure) and people's desire to continue emulating the teachers who had inspired them. It's pointless to argue about whether these issues are "real" or not: faculty believe they are, so any plan to work with faculty needs to address them[4].

[Bark2015] did a two-part study of how computer science educators adopt new teaching practices as individuals, organizationally, and in society as a whole. They asked and answered three key questions:

How do faculty hear about new teaching practices? They intentionally seek out new practices because they're motivated to solve a problem (particularly student engagement), are made aware through deliberate initiatives by their institutions, pick them up from colleagues, or get them from expected *and unexpected* interactions at conferences (teaching-related or otherwise).

Why do they try them out? Sometimes because of institutional incentives (e.g., they innovate to improve their chances of promotion), but there is often tension at research institutions where rhetoric about the importance of teaching is largely disbelieved. Another important reason is their own cost/benefit analysis: will the innovation save them time? A third is that they are inspired by role models— again, this largely affects innovations aimed to improve engagement and motivation rather than learning outcomes—and a fourth is trusted sources, e.g., people they meet at conferences who are in the same situation they are and reported successful adoption.

But faculty had concerns that were often not addressed by people advocating changes. The first was Glass's Law: any new tool or practice initially slows you down, so while new practices might make teaching more effective in the long run, they can't be afforded in the short run. Another is that the physical layout of classrooms makes many new practices hard: for example, discussion groups don't work well in theater-style seating.

---

[3]https://www.insidehighered.com/news/2017/07/06/anthropologist-studies-why-professors-dont-adopt-innovative-teaching-methods

[4]And the prevalence of fixed mindset among faculty when it comes to teaching, i.e., the belief that some people are "just better teachers."

But the most telling result was this: "Despite being researchers themselves, the CS faculty we spoke to for the most part did not believe that results from educational studies were credible reasons to try out teaching practices." This is consistent with other findings: even people whose entire careers are devoted to research often disregard educational research.

Why do they keep using them? As [Bark2015] says, "Student feedback is critical," and is often the strongest reason to continue using a practice, even though we know that learners' self-reports don't correlate strongly with learning outcomes [Star2014; Uttl2017] (though attendance in lectures is a good indicator of engagement). Another reason to retaining a practice is institutional requirements, although if this is the only motivation, people will often drop the practice when the explicit incentive or monitoring is removed.

The good news is that you can tackle these problems systematically. [Baue2015] looked at adoption of new medical techniques within the US Veterans Administration. They found that evidence-based practices in medicine take an average of 17 years to be incorporated into routine general practice, and that only about half of such practices are ever widely adopted. This depressing finding and others like it spurred the growth of **implementation science**, which is the study of how to get people to adopt better practices.

As Chapter 13 said, the starting point is to find out what the people you're trying to help believe they need. For example, [Yada2016] summarizes feedback from K-12 teachers on the preparation and support they want. While it may not all be applicable to all settings, having a cup of tea with a few people and listening before you speak makes a world of difference to their willingness to try something new.

Once you know what people need, the next step is to make changes incrementally, within institutions' own frameworks. [Nara2018] describes an intensive three-year bachelor's program based on tight-knit cohorts and administrative support that tripled graduation rates, while [Hu2017] describes impact of introducing a six-month certification program for existing high school teachers who want to teach computing. The number of computing teachers had been stable from 2007 to 2013, but quadrupled after introduction of the new certification program without diluting quality: new-to-computing teachers seemed to be as effective as teachers with more computing training at teaching the introductory course.

More broadly, [Borr2014] categorizes ways to make change happen in higher education. The categories are defined by whether the change is individual or systemic and whether it is prescribed (top-down) or emergent (bottom-up). The person trying to make the changes (and make them stick) has a different role in each situation, and should pursue different strategies accordingly. The paper goes on to explain each of the methods in detail, while [Hend2015a; Hend2015b] present the same ideas in more actionable form.

Coming from outside, you will probably fall into the Individual/Emergent category to start with, since you will be approaching teachers one by one and trying to make change happen bottom-up. If this is the case, the strategies Borrego and

Henderson recommend center around having teachers reflect on their teaching individually or in groups. Live coding to show them what you do or the examples you use, then having them live code in turn to show how they would use those ideas and techniques in their setting, gives everyone a chance to pick up things that will be useful to them in their context.

14.5 FREE-RANGE TEACHING

Schools and universities aren't the only places people go to learn programming; over the past few years, a growing number have turned to free-range workshops and intensive bootcamp programs. The latter are typically one to six months long, run by volunteer groups or by for-profit companies, and target people who are retraining to get into tech. Some are very high quality, but others exist primarily to separate people from their money [McMi2017].

[Thay2017] interviewed 26 alumni of such bootcamps that provide a second chance for those who missed computing education opportunities earlier (though phrasing it this way makes some pretty big assumptions when it comes to people from underrepresented groups). Bootcamp participants face great personal costs and risks: they must spend significant time, money, and effort before, during, and after bootcamps, and changing careers can take a year or more. Several interviewees felt that their certificates were looked down on by employers; as some said, getting a job means passing an interview, but since interviewers often won't share their reasons for rejection, it's hard to know what to fix or what else to learn. Many resorted to internships (paid or otherwise) and spent a lot of time building their portfolios and networking. The three informal barriers they most clearly identified were jargon, impostor syndrome, and a sense of not fitting in.

[Burk2018] dug into this a bit deeper by comparing the skills and credentials that tech industry recruiters are looking for to those provided by four-year degrees and bootcamps. Based on interviews with 15 hiring managers from firms of various sizes and some focus groups, they found that recruiters uniformly emphasized soft skills (especially teamwork, communication, and the ability to continue learning). Many companies required a four-year degree (though not necessarily in computer science), but many also praised bootcamp graduates for being older or more mature and having more up-to-date knowledge.

If you are approaching an existing bootcamp, your best strategy could be to emphasize what you know about teaching rather than what you know about tech, since many of their founders and staff have programming backgrounds but little or no training in education. The first few chapters of this book have played well with this audience in the past, and [Lang2016] describes evidence-based teaching practices that can be put in place with minimal effort and at low cost. These may not have the most impact, but scoring a few early wins helps build support for larger efforts.

14.6 FINAL THOUGHTS

It is impossible to change large institutions on your own: you need allies, and to get allies, you need tactics. The most useful guide I have found is [Mann2015], which catalogs more than four dozen of these and organizes them according to whether they're best deployed early, later, throughout the change cycle, or when you encounter resistance. A handful of their patterns include:

In Your Space: Keep the new idea visible by placing reminders throughout the organization.

Token: To keep a new idea alive in a person's memory, hand out tokens that can be identified with the topic being introduced.

Champion Skeptic: Ask strong opinion leaders who are skeptical of your new idea to play the role of "official skeptic." Use their comments to improve your effort, even if you don't change their minds.

Future Commitment: If you are able to anticipate some of your needs, you can ask for a future commitment from busy people. If given some lead time, they may be more willing to help.

The most important strategy is to be willing to change your goals based on what you learn from the people you are trying to help. Tutorials showing them how to use a spreadsheet might help them more quickly and more reliably than an introduction to JavaScript. I have often made the mistake of confusing things I was passionate about with things that other people ought to know; if you truly want to be a partner, always remember that learning and change have to go both ways.

The hardest part about building relationships is getting started. Set aside an hour or two every month to find allies and maintain your relationships with them. One way to do this is to ask them for advice: how do they think you ought to raise awareness of what you're doing? Where have they found space to run classes? What needs do they think aren't being met and would you be able to meet them? Any group that has been around for a few years will have useful advice; they will also be flattered to be asked, and will know who you are the next time you call.

And as [Kuch2011] says, if you can't be first in a category, try to create a new category that you can be first in. If you can't do that, join an existing group or think about doing something else entirely. This isn't defeatist: if someone else is already doing what you have in mind, you should either chip in or tackle one of the other equally useful things you could be doing instead.

14.7 EXERCISES

PITCHING A CITY COUNCILOR (INDIVIDUAL/10)

This chapter described an organization that offers weekend programming workshops for people re-entering the workforce. Write an elevator pitch for that organization aimed at a city councilor whose support the organization needs.

PITCHING YOUR ORGANIZATION (INDIVIDUAL/30)

Identify two groups of people your organization needs support from and write an elevator pitch aimed at each one.

EMAIL SUBJECTS (PAIRS/10)

Write the subject lines (and only the subject lines) for three email messages: one announcing a new course, one announcing a new sponsor, and one announcing a change in project leadership. Compare your subject lines to a partner's and see if you can merge the best features of each while also shortening them.

HANDLING PASSIVE RESISTANCE (SMALL GROUPS/30)

People who don't want change will sometimes say so out loud, but may also often use various forms of passive resistance, such as just not getting around to it over and over again, or raising one possible problem after another to make the change seem riskier and more expensive than it's actually likely to be [Scot1987]. Working in small groups, list three or four reasons why people might not want your teaching initiative to go ahead, and explain what you can do with the time and resources you have to counteract each.

WHY LEARN TO PROGRAM? (INDIVIDUAL/15)

Revisit the "Why Learn to Program?" exercise in Section 1.4. Where do your reasons for teaching and your learners' reasons for learning align? Where do they not? How does that affect your marketing?

CONVERSATIONAL PROGRAMMERS (THINK-PAIR-SHARE/15)

A **conversational programmer** is someone who needs to know enough about computing to have a meaningful conversation with a programmer, but isn't going to program themselves. [Wang2018] found that most learning resources don't address this group's needs. Working in pairs, write a pitch for a half-day workshop intended to help people that fit this description and then share your pair's pitch with the rest of the class.

COLLABORATIONS (SMALL GROUPS/30)

Answer the following questions on your own, then compare your answers to those given by other members of your group.

1. Do you have any agreements or relationships with other groups?
2. Do you want to have relationships with any other groups?
3. How would having (or not having) collaborations help you to achieve your goals?
4. What are your key collaborative relationships?

5. Are these the right collaborators for achieving your goals?
6. What groups or entities would you like your organization to have agreements or relationships with?

EDUCATIONALIZATION (WHOLE CLASS/10)

[Laba2008] explores why the United States and other countries keep pushing the solution of social problems onto educational institutions and why that continues not to work. As he points out, "[Education] has done very little to promote equality of race, class, and gender; to enhance public health, economic productivity, and good citizenship; or to reduce teenage sex, traffic deaths, obesity, and environmental destruction. In fact, in many ways it has had a negative effect on these problems by draining money and energy away from social reforms that might have had a more substantial impact." He goes on to write:

> So how are we to understand the success of this institution in light of its failure to do what we asked of it? One way of thinking about this is that education may not be doing what we ask, but it is doing what we want. We want an institution that will pursue our social goals in a way that is in line with the individualism at the heart of the liberal ideal, aiming to solve social problems by seeking to change the hearts, minds, and capacities of individual students. Another way of putting this is that we want an institution through which we can express our social goals without violating the principle of individual choice that lies at the center of the social structure, even if this comes at the cost of failing to achieve these goals. So education can serve as a point of civic pride, a showplace for our ideals, and a medium for engaging in uplifting but ultimately inconsequential disputes about alternative visions of the good life. At the same time, it can also serve as a convenient whipping boy that we can blame for its failure to achieve our highest aspirations for ourselves as a society.

How do efforts to teach computational thinking and digital citizenship in schools fit into this framework? Do bootcamps avoid these traps or simply deliver them in a new guise?

INSTITUTIONAL ADOPTION (WHOLE CLASS/15)

Re-read the list of motivations to adopt new practices given in Section 14.4. Which of these apply to you and your colleagues? Which are irrelevant to your context? Which do you emphasize if and when you interact with people working in formal educational institutions?

IF AT FIRST YOU DON'T SUCCEED (SMALL GROUPS/15)

W.C. Fields probably never said, "If at first you don't succeed, try, try again. Then quit—there's no use being a damn fool about it." It's still good advice: if the people you're trying to reach aren't responding, it could be that you'll never convince them. In groups of 3–4, make a short list of signs that you should stop trying to do something you believe in. How many of them are already true?

MAKING IT FAIL (INDIVIDUAL/15)

[Farm2006] presents some tongue-in-cheek rules for ensuring that new tools *aren't* adopted, all of which apply to new teaching practices:

1. Make it optional.
2. Economize on training.
3. Don't use it in a real project.
4. Never integrate it.
5. Use it sporadically.
6. Make it part of a quality initiative.
7. Marginalize the champion.
8. Capitalize on early missteps.
9. Make a small investment.
10. Exploit fear, uncertainty, doubt, laziness, and inertia.

Which of these have you seen done recently? Which have you done yourself? What form did they take?

MENTORING (WHOLE CLASS/15)

The Institute for African-American Mentoring in Computer Science[5] has published guidelines for mentoring doctoral students[6]. Read them individually, then go through them as a class and rate your efforts for your own group as +1 (definitely doing), 0 (not sure or not applicable), or -1 (definitely not doing).

---

[5]http://www.iaamcs.org/

[6]http://iaamcs.org/guidelines

15 Why I Teach

When I started volunteering at the University of Toronto, some of my students asked me why I would teach for free. This was my answer:

> When I was your age, I thought universities existed to teach people how to learn. Later, in grad school, I thought universities were about doing research and creating new knowledge. Now that I'm in my forties, though, I've realized that what we're really teaching you is how to take over the world, because you're going to have to whether you want to or not.
>
> My parents are in their seventies. They don't run the world any more; it's people my age who pass laws and make life-and-death decisions in hospitals. As scary as it is, *we* are the grownups.
>
> Twenty years from now, we'll be heading for retirement and *you* will be in charge. That may sound like a long time when you're nineteen, but take three breaths and it's gone. That's why we give you problems whose answers can't be cribbed from last year's notes. That's why we put you in situations where you have to figure out what needs to be done right now, what can be left for later, and what you can simply ignore. It's because if you don't learn how to do these things now, you won't be ready to do them when you have to.

It was all true, but it wasn't the whole story. I don't want people to make the world a better place so that I can retire in comfort. I want them to do it because it's the greatest adventure of our time. A hundred and fifty years ago, most societies practiced slavery. A hundred years ago, my grandmother wasn't legally a person[1] in Canada. In the year I was born, most of the world's people suffered under totalitarian rule, and judges were still ordering electroshock therapy to "cure" homosexuals. There's still a lot wrong with the world, but look at how many more choices we have than our grandparents did. Look at how many more things we can know, and be, and enjoy because we're finally taking the Golden Rule seriously.

I am less optimistic today than I was then. Climate change, mass extinction, surveillance capitalism, inequality on a scale we haven't seen in a century, the re-emergence of racist nationalism: my generation has watched it all happen and shrugged. The bill for our cowardice, lethargy, and greed won't come due until my daughter is grown, but it *will* come, and by the time it does there will be no easy solutions to these problems (and possibly no solutions at all).

So this is why I teach today: I'm angry. I'm angry because your sex and your color and your parents' wealth and connections shouldn't count for more than how smart or honest or hard-working you are. I'm angry because we turned the Internet into a

---

[1] https://en.wikipedia.org/wiki/The_Famous_Five_(Canada)

cesspool. I'm angry because Nazis are on the march once again and billionaires are playing with rocket ships while the planet is melting. I'm angry, so I teach, because the world only gets better when we teach people how to make it better.

In his 1947 essay "Why I Write", George Orwell[2] wrote:

> In a peaceful age I might have written ornate or merely descriptive books, and might have remained almost unaware of my political loyalties. As it is I have been forced into becoming a sort of pamphleteer... Every line of serious work that I have written since 1936 has been written, directly or indirectly, against totalitarianism... It seems to me nonsense, in a period like our own, to think that one can avoid writing of such subjects. Everyone writes of them in one guise or another. It is simply a question of which side one takes.

Replace "writing" with "teaching" and you'll have the reason I do what I do.

Thank you for reading. I hope we can learn something together some day. Until then, please:

Start where you are.
Use what you have.
Help who you can.

---

[2]http://www.resort.com/ prime8/Orwell/whywrite.html

Bibliography

[Abba2012] Janet Abbate. "Recoding Gender: Women's Changing Participation in Computing." Describes the careers and accomplishments of the women who shaped the early history of computing, but have all too often been written out of that history. MIT Press, 2012. ISBN: 9780262534536.

[Abel2009] Andrew Abela. *Chart Suggestions - A Thought Starter*. http : / / extremepresentation.typepad.com/files/choosing-a-good-chart-09.pdf. A graphical decision tree for choosing the right type of chart. 2009.

[Adam1975] Frank Adams and Myles Horton. "Unearthing Seeds of Fire: The Idea of Highlander." A history of the Highlander Folk School and its founder, Myles Horton. Blair, 1975. ISBN: 0895870193.

[Aike1975] Edwin G. Aiken, Gary S. Thomas, and William A. Shennum. "Memory for a Lecture: Effects of Notes, Lecture Rate, and Informational Density." In: *Journal of Educational Psychology* 67.3 (1975). An early landmark study showing that taking notes improved retention, pp. 439–444. DOI: 10.1037/h0076613.

[Aiva2016] Efthimia Aivaloglou and Felienne Hermans. "How Kids Code and How We Know." In: *2016 International Computing Education Research Conference (ICER'16)*. Presents an analysis of 250,000 Scratch projects. Association for Computing Machinery (ACM), 2016. DOI: 10.1145/2960310.2960325.

[Alin1989] Saul D. Alinsky. "Rules for Radicals: A Practical Primer for Realistic Radicals." A widely-read guide to community organization written by one of the 20th century's great organizers. Vintage, 1989. ISBN: 0679721134.

[Alqa2017] Basma S. Alqadi and Jonathan I. Maletic. "An Empirical Study of Debugging Patterns Among Novice Programmers." In: *2017 Technical Symposium on Computer Science Education (SIGCSE'17)*. Reports patterns in the debugging activities and success rates of novice programmers. Association for Computing Machinery (ACM), 2017. DOI: 10.1145/3017680.3017761.

[Alvi1999] Jennifer Alvidrez and Rhona S. Weinstein. "Early Teacher Perceptions and Later Student Academic Achievement." In: *Journal of Educational Psychology* 91.4 (1999). An influential study of the effects of teachers' perceptions of students on their later achievements, pp. 731–746. DOI: 10.1037/0022-0663.91.4.731.

[Ambr2010] Susan A. Ambrose, Michael W. Bridges, Michele DiPietro, Marsha C. Lovett, and Marie K. Norman. "How Learning Works: Seven Research-Based Principles for Smart Teaching." Summarizes what we know about education and why we believe it's true, from cognitive psychology to social factors. Jossey-Bass, 2010. ISBN: 0470484101.

[Ande2001] Lorin W. Anderson and David R. Krathwohl, eds. "A Taxonomy for Learning, Teaching, And Assessing: A Revision of Bloom's Taxonomy of Educational Objectives." A widely used revision to Bloom's Taxonomy. Longman, 2001. ISBN: 080131903X.

[Armo2008] Michal Armoni and David Ginat. "Reversing: A Fundamental Idea in Computer Science." In: *Computer Science Education* 18.3 (Sept. 2008). Argues that the notion of reversing things is an unrecognized fundamental concept in computing education, pp. 213–230. DOI: 10 . 1080 / 08993400802332670.

[Atki2000] Robert K. Atkinson, Sharon J. Derry, Alexander Renkl, and Donald Wortham. "Learning from Examples: Instructional Principles from the Worked Examples Research." In: *Review of Educational Research* 70.2 (June 2000). A comprehensive survey of worked examples research at the time, pp. 181–214. DOI: 10.3102/00346543070002181.

[Auro2019] Valerie Aurora and Mary Gardiner. "How to Respond to Code of Conduct Reports." Version 1.1. A short, practical guide to enforcing a Code of Conduct. Frame Shift Consulting LLC, 2019. ISBN: 978-1386922575.

[Avel2013] Emma-Louise Aveling, Peter McCulloch, and Mary Dixon-Woods. "A Qualitative Study Comparing Experiences of the Surgical Safety Checklist in Hospitals in High-Income and Low-Income Countries." In: *BMJ Open* 3.8 (Aug. 2013). Reports the effectiveness of surgical checklist implementations in the UK and Africa. DOI: 10.1136/bmjopen-2013-003039.

[Bacc2013] Alberto Bacchelli and Christian Bird. "Expectations, Outcomes, and Challenges of Modern Code Review." In: *2013 International Conference on Software Engineering (ICSE'13)*. A summary of work on code review. May 2013.

[Bari2017] Titus Barik, Justin Smith, Kevin Lubick, Elisabeth Holmes, Jing Feng, Emerson Murphy-Hill, and Chris Parnin. "Do Developers Read Compiler Error Messages?." In: *2017 International Conference on Software Engineering (ICSE'17)*. Reports that developers do read error messages and doing so is as hard as reading source code: it takes 13-25% of total task time. Institute of Electrical and Electronics Engineers (IEEE), May 2017. DOI: 10.1109/icse. 2017.59.

[Bark2014] Lecia Barker, Christopher Lynnly Hovey, and Leisa D. Thompson. "Results of a Large-Scale, Multi-Institutional Study of Undergraduate Retention in Computing." In: *2014 Frontiers in Education Conference (FIE'14)*. Reports that meaningful assignments, faculty interaction with students, student collaboration on assignments, and (for male students) pace and workload relative to expectations drive retention in computing classes, while interactions with teaching assistants or with peers in extracurricular activities have little impact. Institute of Electrical and Electronics Engineers (IEEE), Oct. 2014. DOI: 10.1109/fie.2014.7044267.

[**Bark2015**] Lecia Barker, Christopher Lynnly Hovey, and Jane Gruning. "What Influences CS Faculty to Adopt Teaching Practices?." In: *2015 Technical Symposium on Computer Science Education (SIGCSE'15)*. Describes how computer science educators adopt new teaching practices. Association for Computing Machinery (ACM), 2015. DOI: 10.1145/2676723.2677282.

[**Basi1987**] Victor R. Basili and Richard W. Selby. "Comparing the Effectiveness of Software Testing Strategies." In: *IEEE Transactions on Software Engineering* SE-13.12 (Dec. 1987). An early and influential summary of the effectiveness of code review, pp. 1278–1296. DOI: 10.1109/tse.1987.232881.

[**Basu2015**] Soumya Basu, Albert Wu, Brian Hou, and John DeNero. "Problems Before Solutions: Automated Problem Clarification at Scale." In: *2015 Conference on Learning @ Scale (L@S'15)*. Describes a system in which students have to unlock test cases for their code by answering MCQs, and presents data showing that this is effective. Association for Computing Machinery (ACM), 2015. DOI: 10.1145/2724660.2724679.

[**Batt2018**] Lina Battestilli, Apeksha Awasthi, and Yingjun Cao. "Two-Stage Programming Projects: Individual Work Followed by Peer Collaboration." In: *2018 Technical Symposium on Computer Science Education (SIGCSE'18)*. Reports that learning outcomes were improved by two-stage projects in which students work individually, then re-work the same problem in pairs. Association for Computing Machinery (ACM), 2018. DOI: 10.1145/3159450.3159486.

[**Baue2015**] Mark S. Bauer, Laura Damschroder, Hildi Hagedorn, Jeffrey Smith, and Amy M. Kilbourne. "An Introduction to Implementation Science for the Non-Specialist." In: *BMC Psychology* 3.1 (Sept. 2015). Explains what implementation science is, using examples from the US Veterans Administration to illustrate. DOI: 10.1186/s40359-015-0089-9.

[**Beck2013**] Leland Beck and Alexander Chizhik. "Cooperative Learning Instructional methods for CS1: Design, Implementation, and Evaluation." In: *ACM Transactions on Computing Education* 13.3 (Aug. 2013). Reports that cooperative learning enhances learning outcomes and self-efficacy in CS1, 10:1–10:21. ISSN: 1946-6226. DOI: 10.1145/2492686.

[**Beck2014**] Victoria Beck. "Testing a Model to Predict Online Cheating—Much Ado About Nothing." In: *Active Learning in Higher Education* 15.1 (Jan. 2014). Reports that cheating is no more likely in online courses than in face-to-face courses, pp. 65–75. DOI: 10.1177/1469787413514646.

[**Beck2016**] Brett A. Becker, Graham Glanville, Ricardo Iwashima, Claire McDonnell, Kyle Goslin, and Catherine Mooney. "Effective Compiler Error Message Enhancement for Novice Programming Students." In: *Computer Science Education* 26.2-3 (July 2016). Reports that improved error messages helped novices learn faster, pp. 148–175. DOI: 10.1080/08993408.2016.1225464.

[**Beni2017**] Gal Beniamini, Sarah Gingichashvili, Alon Klein Orbach, and Dror G. Feitelson. "Meaningful Identifier Names: The Case of Single-Letter

Variables." In: *2017 International Conference on Program Comprehension (ICPC'17)*. Reports that use of single-letter variable names doesn't affect ability to modify code, and that some single-letter variable names have implicit types and meanings. Institute of Electrical and Electronics Engineers (IEEE), May 2017. DOI: 10.1109/icpc.2017.18.

[**Benn2000**] Patricia Benner. "From Novice to Expert: Excellence and Power in Clinical Nursing Practice." A classic study of clinical judgment and the development of expertise. Pearson, 2000. ISBN: 0130325228.

[**Benn2007a**] Jens Bennedsen and Michael E. Caspersen. "Failure Rates in Introductory Programming." In: *ACM SIGCSE Bulletin* 39.2 (June 2007). Reports that 67% of students pass CS1, with variation from 5% to 100%, p. 32. DOI: 10.1145/1272848.1272879.

[**Benn2007b**] Jens Bennedsen and Carsten Schulte. "What Does "Objects-first" Mean?: An International Study of Teachers' Perceptions of Objects-first." In: *2007 Koli Calling Conference on Computing Education Research (Koli'07)*. Teases out three meanings of "objects first" in computing education. 2007, pp. 21–29.

[**Berg2012**] Joseph Bergin, Jane Chandler, Jutta Eckstein, Helen Sharp, Mary Lynn Manns, Klaus Marquardt, Marianna Sipos, Markus Völter, and Eugene Wallingford. "Pedagogical Patterns: Advice for Educators." A catalog of design patterns for teaching. CreateSpace, 2012. ISBN: 9781479171828.

[**Biel1995**] Katerine Bielaczyc, Peter L. Pirolli, and Ann L. Brown. "Training in Self-Explanation and Self-Regulation Strategies: Investigating the Effects of Knowledge Acquisition Activities on Problem Solving." In: *Cognition and Instruction* 13.2 (June 1995). Reports that training learners in self-explanation accelerates their learning, pp. 221–252. DOI: 10.1207/s1532690xci1302_3.

[**Bigg2011**] John Biggs and Catherine Tang. "Teaching for Quality Learning at University." A step-by-step guide to lesson development, delivery, and evaluation for people working in higher education. Open University Press, 2011. ISBN: 0335242758.

[**Bink2012**] Dave Binkley, Marcia Davis, Dawn Lawrie, Jonathan I. Maletic, Christopher Morrell, and Bonita Sharif. "The Impact of Identifier Style on Effort and Comprehension." In: *Empirical Software Engineering* 18.2 (May 2012). Reports that reading and understanding code is fundamentally different from reading prose, and that experienced developers are relatively unaffected by identifier style, but beginners benefit from the use of camel case (versus pothole case), pp. 219–276. DOI: 10.1007/s10664-012-9201-4.

[**Blik2014**] Paulo Blikstein, Marcelo Worsley, Chris Piech, Mehran Sahami, Steven Cooper, and Daphne Koller. "Programming Pluralism: Using Learning Analytics to Detect Patterns in the Learning of Computer Programming." In: *Journal of the Learning Sciences* 23.4 (Oct. 2014). Reports an attempt to categorize novice programmer behavior using machine learning that found interesting patterns on individual assignments, pp. 561–599. DOI: 10.1080/10508406. 2014.954750.

[Bloo1984] Benjamin S. Bloom. "The 2 Sigma Problem: The Search for Methods of Group Instruction as Effective as One-to-One Tutoring." In: *Educational Researcher* 13.6 (June 1984). Reports that students tutored one-to-one using mastery learning techniques perform two standard deviations better than those who learned through conventional lecture, pp. 4–16. DOI: 10.3102/0013189x013006004.

[Boha2011] Mark Bohay, Daniel P. Blakely, Andrea K. Tamplin, and Gabriel A. Radvansky. "Note Taking, Review, Memory, and Comprehension." In: *American Journal of Psychology* 124.1 (2011). Reports that note-taking improves retention most at deeper levels of understanding, p. 63. DOI: 10.5406/amerjpsyc.124.1.0063.

[Boll2014] David Bollier. "Think Like a Commoner: A Short Introduction to the Life of the Commons." A short introduction to a widely-used model of governance. New Society Publishers, 2014. ISBN: 0865717680.

[Borr2014] Maura Borrego and Charles Henderson. "Increasing the Use of Evidence-Based Teaching in STEM Higher Education: A Comparison of Eight Change Strategies." In: *Journal of Engineering Education* 103.2 (Apr. 2014). Categorizes different approaches to effecting change in higher education, pp. 220–252. DOI: 10.1002/jee.20040.

[Bria2015] Samuel A. Brian, Richard N. Thomas, James M. Hogan, and Colin Fidge. "Planting Bugs: A System for Testing Students' Unit Tests." In: *2015 Conference on Innovation and Technology in Computer Science Education (ITiCSE'15)*. Describes a tool for assessing students' programs and unit tests and finds that students often write weak tests and misunderstand the role of unit testing. Association for Computing Machinery (ACM), 2015. DOI: 10.1145/2729094.2742631.

[Broo2016] Stephen D. Brookfield and Stephen Preskill. "The Discussion Book: 50 Great Ways to Get People Talking." Describes fifty different ways to get groups talking productively. Jossey-Bass, 2016. ISBN: 9781119049715.

[Brop1983] Jere E. Brophy. "Research on the Self-Fulfilling Prophecy and Teacher Expectations." In: *Journal of Educational Psychology* 75.5 (1983). A early, influential study of the effects of teachers' perceptions on students' achievements, pp. 631–661. DOI: 10.1037/0022-0663.75.5.631.

[Brow2007] Michael Jacoby Brown. "Building Powerful Community Organizations: A Personal Guide to Creating Groups that Can Solve Problems and Change the World." A practical guide to creating effective organizations in and for communities. Long Haul Press, 2007. ISBN: 0977151808.

[Brow2017] Neil C. C. Brown and Amjad Altadmri. "Novice Java Programming Mistakes." In: *ACM Transactions on Computing Education* 17.2 (May 2017).

Summarizes the authors' analysis of novice programming mistakes. DOI: 10.1145/2994154.

[Brow2018] Neil C. C. Brown and Greg Wilson. "Ten Quick Tips for Teaching Programming." In: *PLoS Computational Biology* 14.4 (Apr. 2018). A short summary of what we actually know about teaching programming and why we believe it's true. DOI: 10.1371/journal.pcbi.1006023.

[Buff2015] Kevin Buffardi and Stephen H. Edwards. "Reconsidering Automated Feedback: A Test-Driven Approach." In: *2015 Technical Symposium on Computer Science Education (SIGCSE'15)*. Describes a system that associates failed tests with particular features in a learner's code so that learners cannot game the system. Association for Computing Machinery (ACM), 2015. DOI: 10.1145/2676723.2677313.

[Burg2015] Sheryl E. Burgstahler. "Universal Design in Higher Education: From Principles to Practice." Second. Describes how to make online teaching materials accessible to everyone. Harvard Education Press, 2015. ISBN: 9781612508160.

[Burk2018] Quinn Burke, Cinamon Bailey, Louise Ann Lyon, and Emily Greeen. "Understanding the Software Development Industry's Perspective on Coding Boot Camps Versus Traditional 4-Year Colleges." In: *2018 Technical Symposium on Computer Science Education (SIGCSE'18)*. Compares the skills and credentials that tech industry recruiters are looking for to those provided by 4-year degrees and bootcamps. Association for Computing Machinery (ACM), 2018. DOI: 10.1145/3159450.3159485.

[Butl2017] Zack Butler, Ivona Bezakova, and Kimberly Fluet. "Pencil Puzzles for Introductory Computer Science." In: *2017 Technical Symposium on Computer Science Education (SIGCSE'17)*. Describes pencil-and-paper puzzles that can be turned into CS1/CS2 assignments, and reports that they are enjoyed by students and encourage meta-cognition. Association for Computing Machinery (ACM), 2017. DOI: 10.1145/3017680.3017765.

[Byck2005] Pauli Byckling, Petri Gerdt, and Jorma Sajaniemi. "Roles of Variables in Object-Oriented Programming." In: *2005 Conference on Object-Oriented Programming, Systems, Languages, and Applications (OOPSLA'05)*. Presents single-variable design patterns common in novice programs. Association for Computing Machinery (ACM), 2005. DOI: 10.1145/1094855.1094972.

[Camp2016] Jennifer Campbell, Diane Horton, and Michelle Craig. "Factors for Success in Online CS1." In: *2016 Conference on Innovation and Technology in Computer Science Education (ITiCSE'16)*. Compares students who opted into an online CS1 class online with those who took it in person in a flipped classroom. Association for Computing Machinery (ACM), 2016. DOI: 10.1145/2899415.2899457.

[Carr1987] John Carroll, Penny Smith-Kerker, James Ford, and Sandra Mazur-Rimetz. "The Minimal Manual." In: *Human-Computer Interaction* 3.2 (June

1987). The foundational paper on minimalist instruction, pp. 123–153. DOI: 10.1207/s15327051hci0302_2.

[**Carr2014**] John Carroll. "Creating Minimalist Instruction." In: *International Journal of Designs for Learning* 5.2 (Nov. 2014). A look back on the author's work on minimalist instruction. DOI: 10.14434/ijdl.v5i2.12887.

[**Cart2017**] Adam Scott Carter and Christopher David Hundhausen. "Using Programming Process Data to Detect Differences in Students' Patterns of Programming." In: *2017 Technical Symposium on Computer Science Education (SIGCSE'17)*. Shows that students of different levels approach programming tasks differently, and that these differences can be detected automatically. Association for Computing Machinery (ACM), 2017. DOI: 10.1145/3017680.3017785.

[**Casp2007**] Michael E. Caspersen and Jens Bennedsen. "Instructional Design of a Programming Course." In: *2007 International Computing Education Research Conference (ICER'07)*. Goes from a model of human cognition to three learning theories, and from there to the design of an introductory object-oriented programming course. Association for Computing Machinery (ACM), 2007. DOI: 10.1145/1288580.1288595.

[**Cele2018**] Mehmet Celepkolu and Kristy Elizabeth Boyer. "Thematic Analysis of Students' Reflections on Pair Programming in CS1." In: *2018 Technical Symposium on Computer Science Education (SIGCSE'18)*. Reports that pair programming has the same learning gains side-by-side programming but higher student satisfaction. Association for Computing Machinery (ACM), 2018. DOI: 10.1145/3159450.3159516.

[**Ceti2016**] Ibrahim Cetin and Christine Andrews-Larson. "Learning Sorting Algorithms Through Visualization Construction." In: *Computer Science Education* 26.1 (Jan. 2016). Reports that people learn more from constructing algorithm visualizations than they do from viewing visualizations constructed by others, pp. 27–43. DOI: 10.1080/08993408.2016.1160664.

[**Chen2009**] Nicholas Chen and Maurice Rabb. "A Pattern Language for Screencasting." In: *2009 Conference on Pattern Languages of Programs (PLoP'09)*. A brief, well-organized collection of tips for making screencasts. Association for Computing Machinery (ACM), 2009. DOI: 10.1145/1943226.1943234.

[**Chen2017**] Nick Cheng and Brian Harrington. "The Code Mangler: Evaluating Coding Ability Without Writing Any Code." In: *2017 Technical Symposium on Computer Science Education (SIGCSE'17)*. Reports that student performance on exercises in which they undo code mangling correlates strongly with performance on traditional assessments. Association for Computing Machinery (ACM), 2017. DOI: 10.1145/3017680.3017704.

[**Chen2018**] Chen Chen, Paulina Haduong, Karen Brennan, Gerhard Sonnert, and Philip Sadler. "The effects of first programming language on college students' computing attitude and achievement: a comparison of graphical and textual languages." In: *Computer Science Education* 29.1 (Nov. 2018). Finds that students whose first language was graphical had higher grades than students whose first language was textual when the languages were introduced in or before early adolescent years, pp. 23–48. DOI: 10.1080/08993408.2018.1547564. URL: https://doi.org/10.1080/08993408.2018.1547564.

[**Cher2007**] Mauro Cherubini, Gina Venolia, Rob DeLine, and Andrew J. Ko. "Let's Go to the Whiteboard: How and Why Software Developers Use Drawings." In: *2007 Conference on Human Factors in Computing Systems (CHI'07)*. Reports that developers draw diagrams to aid discussion rather than to document designs. Association for Computing Machinery (ACM), 2007. DOI: 10.1145/1240624.1240714.

[**Cher2009**] Sapna Cheryan, Victoria C. Plaut, Paul G. Davies, and Claude M. Steele. "Ambient Belonging: How Stereotypical Cues Impact Gender Participation in Computer Science." In: *Journal of Personality and Social Psychology* 97.6 (2009). Reports that subtle environmental clues have a measurable impact on the interest that people of different genders have in computing, pp. 1045–1060. DOI: 10.1037/a0016239.

[**Chet2014**] Raj Chetty, John N. Friedman, and Jonah E. Rockoff. "Measuring the Impacts of Teachers II: Teacher Value-Added and Student Outcomes in Adulthood." In: *American Economic Review* 104.9 (Sept. 2014). Reports that good teachers have a small but measurable impact on student outcomes, pp. 2633–2679. DOI: 10.1257/aer.104.9.2633.

[**Chi1989**] Michelene T. H. Chi, Miriam Bassok, Matthew W. Lewis, Peter Reimann, and Robert Glaser. "Self-Explanations: How Students Study and Use Examples in Learning to Solve Problems." In: *Cognitive Science* 13.2 (Apr. 1989). A seminal paper on the power of self-explanation, pp. 145–182. DOI: 10.1207/s15516709cog1302_1.

[**Coco2018**] Center for Community Organizations. *The "Problem" Woman of Colour in the Workplace.* https://coco-net.org/problem-woman-colour-nonprofit-organizations/. Outlines the experience of many women of color in the workplace. 2018.

[**Coll1991**] Allan Collins, John Seely Brown, and Ann Holum. "Cognitive Apprenticeship: Making Thinking Visible." In: *American Educator* 6 (1991). Describes an educational model based on the notion of apprenticeship and master guidance, pp. 38–46.

[**Coom2012**] Norman Coombs. "Making Online Teaching Accessible." An accessible guide to making online lessons accessible. Jossey-Bass, 2012. ISBN: 9781458725288.

[**Covi2017**] Martin V. Covington, Linda M. von Hoene, and Dominic J. Voge. "Life Beyond Grades: Designing College Courses to Promote Intrinsic Motivation." Explores ways of balancing intrinsic and extrinsic motivation in institutional education. Cambridge University Press, 2017. ISBN: 9780521805230.

[Craw2010] Matthew B. Crawford. "Shop Class as Soulcraft: An Inquiry into the Value of Work." A deep analysis of what we learn about ourselves by doing certain kinds of work. Penguin, 2010. ISBN: 9780143117469.

[Crou2001] Catherine H. Crouch and Eric Mazur. "Peer Instruction: Ten Years of Experience and Results." In: *American Journal of Physics* 69.9 (Sept. 2001). Reports results from the first ten years of peer instruction in undergraduate physics classes, and describes ways in which its implementation changed during that time, pp. 970–977. DOI: 10.1119/1.1374249.

[Csik2008] Mihaly Csikszentmihaly. "Flow: The Psychology of Optimal Experience." An influential discussion of what it means to be fully immersed in a task. Harper, 2008. ISBN: 978-0061339202.

[Cunn2017] Kathryn Cunningham, Sarah Blanchard, Barbara J. Ericson, and Mark Guzdial. "Using Tracing and Sketching to Solve Programming Problems." In: *2017 Conference on International Computing Education Research (ICER'17)*. Found that writing new values near variables' names as they change is the most effective tracing technique. Association for Computing Machinery (ACM), 2017. DOI: 10.1145/3105726.3106190.

[Cutt2017] Quintin Cutts, Charles Riedesel, Elizabeth Patitsas, Elizabeth Cole, Peter Donaldson, Bedour Alshaigy, Mirela Gutica, Arto Hellas, Edurne Larraza-Mendiluze, and Robert McCartney. "Early Developmental Activities and Computing Proficiency." In: *2017 Conference on Innovation and Technology in Computer Science Education (ITiCSE'17)*. Surveyed adult computer users about childhood activities and found strong correlation between confidence and computer use based on reading on one's own and playing with construction toys with no moving parts (like Lego). Association for Computing Machinery (ACM), 2017. DOI: 10.1145/3174781.3174789.

[Dage2010] Barthélémy Dagenais, Harold Ossher, Rachel K. E. Bellamy, Martin P. Robillard, and Jacqueline P. de Vries. "Moving Into a New Software Project Landscape." In: *2010 International Conference on Software Engineering (ICSE'10)*. A look at how people move from one project or domain to another. ACM Press, 2010. DOI: 10.1145/1806799.1806842.

[Dahl2018] Sarah Dahlby Albright, Titus H. Klinge, and Samuel A. Rebelsky. "A Functional Approach to Data Science in CS1." In: *2018 Technical Symposium on Computer Science Education (SIGCSE'18)*. Describes the design of a CS1 class built around data science. Association for Computing Machinery (ACM), 2018. DOI: 10.1145/3159450.3159550.

[Deb2018] Debzani Deb, Muztaba Fuad, James Etim, and Clay Gloster. "MRS: Automated Assessment of Interactive Classroom Exercises." In: *2018 Technical Symposium on Computer Science Education (SIGCSE'18)*. Reports that doing in-class exercises with realtime feedback using mobile devices improved concept retention and student engagement while reducing failure rates. Association for Computing Machinery (ACM), 2018. DOI: 10.1145/3159450.3159607.

[DeBr2015] Pedro De Bruyckere, Paul A. Kirschner, and Casper D. Hulshof. "Urban Myths about Learning and Education." Describes and debunks some widely held myths about how people learn. Academic Press, 2015. ISBN: 9780128015377.

[Denn2019] Paul Denny, Brett A. Becker, Michelle Craig, Greg Wilson, and Piotr Banaszkiewicz. "Research This! Questions that Computing Educators Most Want Computing Education Researchers to Answer." In: (2019). Found little overlap between the questions that computing education researchers are most interested in and the questions practitioners want answered.

[Derb2006] Esther Derby and Diana Larsen. "Agile Retrospectives: Making Good Teams Great." Describes how to run a good project retrospective. Pragmatic Bookshelf, 2006. ISBN: 0977616649.

[Deve2018] Gabriel A. Devenyi, Rémi Emonet, Rayna M. Harris, Kate L. Hertweck, Damien Irving, Ian Milligan, and Greg Wilson. "Ten Simple Rules for Collaborative Lesson Development." In: *PLoS Computational Biology* 14.3 (Mar. 2018). Describes how to develop lessons together. DOI: 10.1371/journal.pcbi. 1005963.

[Dida2016] David Didau and Nick Rose. "What Every Teacher Needs to Know About Psychology." An informative, opinionated explanation of what modern psychology has to say about teaching. John Catt Educational, 2016. ISBN: 1909717851.

[DiSa2014a] Betsy DiSalvo, Mark Guzdial, Amy Bruckman, and Tom McKlin. "Saving Face While Geeking Out: Video Game Testing as a Justification for Learning Computer Science." In: *Journal of the Learning Sciences* 23.3 (July 2014). Found that 65% of male African-American participants in a game testing program went on to study computing, pp. 272–315. DOI: 10.1080/ 10508406.2014.893434.

[DiSa2014b] Betsy DiSalvo, Cecili Reid, and Parisa Khanipour Roshan. "They Can't Find Us." In: *2014 Technical Symposium on Computer Science Education (SIGCSE'14)*. Reports that the search terms parents were likely to use for out-of-school CS classes didn't actually find those classes. Association for Computing Machinery (ACM), 2014. DOI: 10.1145/2538862.2538933.

[Douc2005] Christopher Douce, David Livingstone, and James Orwell. "Automatic Test-Based Assessment of Programming." In: *Journal on Educational Resources in Computing* 5.3 (Sept. 2005). Reviews the state of auto-graders at the time. DOI: 10.1145/1163405.1163409.

[DuBo1986] Benedict Du Boulay. "Some Difficulties of Learning to Program." In: *Journal of Educational Computing Research* 2.1 (Feb. 1986). Introduces the idea of a notional machine, pp. 57–73. DOI: 10.2190/3lfx-9rrf-67t8-uvk9.

[Edwa2014a] Stephen H. Edwards, Zalia Shams, and Craig Estep. "Adaptively Identifying Non-Terminating Code when Testing Student Programs." In: *2014 Technical Symposium on Computer Science Education (SIGCSE'14)*.

Describes an adaptive scheme for detecting non-terminating student coding submissions. Association for Computing Machinery (ACM), 2014. DOI: 10. 1145/2538862.2538926.

[Edwa2014b] Stephen H. Edwards and Zalia Shams. "Do Student Programmers All Tend to Write the Same Software Tests?." In: *2014 Conference on Innovation and Technology in Computer Science Education (ITiCSE'14)*. Reports that students wrote tests for the happy path rather than to detect hidden bugs. Association for Computing Machinery (ACM), 2014. DOI: 10.1145/2591708.2591757.

[Endr2014] Stefan Endrikat, Stefan Hanenberg, Romain Robbes, and Andreas Stefik. "How Do API Documentation and Static Typing Affect API Usability?." In: *2014 International Conference on Software Engineering (ICSE'14)*. Shows that types do add complexity to programs, but it pays off fairly quickly by acting as documentation hints for a method's use. ACM Press, 2014. DOI: 10.1145/2568225.2568299.

[Ensm2003] Nathan L. Ensmenger. "Letting the "Computer Boys" Take Over: Technology and the Politics of Organizational Transformation." In: *International Review of Social History* 48.S11 (Dec. 2003). Describes how programming was turned from a female into a male profession in the 1960s, pp. 153–180. DOI: 10.1017/s0020859003001305.

[Ensm2012] Nathan L. Ensmenger. "The Computer Boys Take Over: Computers, Programmers, and the Politics of Technical Expertise." Traces the emergence and rise of computer experts in the 20th century, and particularly the way that computing became male-gendered. MIT Press, 2012. ISBN: 9780262517966.

[Eppl2006] Martin J. Eppler. "A Comparison Between Concept Maps, Mind Maps, Conceptual Diagrams, and Visual Metaphors as Complementary Tools for Knowledge Construction and Sharing." In: *Information Visualization* 5.3 (June 2006). Compares concept maps, mind maps, conceptual diagrams, and visual metaphors as learning tools, pp. 202–210. DOI: 10.1057/palgrave.ivs.9500131.

[Epst2002] Lewis Carroll Epstein. "Thinking Physics: Understandable Practical Reality." An entertaining problem-based introduction to thinking like a physicist. Insight Press, 2002. ISBN: 0935218084.

[Eric2015] Barbara J. Ericson, Steven Moore, Briana B. Morrison, and Mark Guzdial. "Usability and Usage of Interactive Features in an Online Ebook for CS Teachers." In: *2015 Workshop in Primary and Secondary Computing Education (WiPSCE'15)*. Reports that learners are more likely to attempt Parsons Problems than nearby multiple choice questions in an ebook. Association for Computing Machinery (ACM), 2015, pp. 111–120. ISBN: 978-1-4503-3753-3. DOI: 10.1145/2818314.2818335.

[Eric2016] K. Anders Ericsson. "Summing Up Hours of Any Type of Practice Versus Identifying Optimal Practice Activities." In: *Perspectives on Psychological Science* 11.3 (May 2016). A critique of a meta-study of deliberate practice based on the latter's overly-broad inclusion of activities, pp. 351–354. DOI: 10.1177/1745691616635600.

[**Eric2017**] Barbara J. Ericson, Lauren E. Margulieux, and Jochen Rick. "Solving Parsons Problems versus Fixing and Writing Code." In: *2017 Koli Calling Conference on Computing Education Research (Koli'17)*. Reports that solving 2D Parsons Problems with distractors takes less time than writing or fixing code but has equivalent learning outcomes. Association for Computing Machinery (ACM), 2017. DOI: 10.1145/3141880.3141895.

[**Farm2006**] Eugene Farmer. "The Gatekeeper's Guide, or How to Kill a Tool." In: *IEEE Software* 23.6 (Nov. 2006). Ten tongue-in-cheek rules for making sure that a new software tool doesn't get adopted, pp. 12–13. DOI: 10.1109/ms. 2006.174.

[**Fehi2008**] Chris Fehily. "SQL: Visual QuickStart Guide." Third. An introduction to SQL that is both a good tutorial and a good reference guide. Peachpit Press, 2008. ISBN: 0321553578.

[**Finc2007**] Sally Fincher and Josh Tenenberg. "Warren's Question." In: *2007 International Computing Education Research Conference (ICER'07)*. A detailed look at a particular instance of transferring a teaching practice. Association for Computing Machinery (ACM), 2007. DOI: 10.1145/1288580.1288588.

[**Finc2012**] Sally Fincher, Brad Richards, Janet Finlay, Helen Sharp, and Isobel Falconer. "Stories of Change: How Educators Change Their Practice." In: *2012 Frontiers in Education Conference (FIE'12)*. A detailed look at how educators actually adopt new teaching practices. Institute of Electrical and Electronics Engineers (IEEE), Oct. 2012. DOI: 10.1109/fie.2012.6462317.

[**Finc2019**] Sally Fincher and Anthony Robins, eds. "The Cambridge Handbook of Computing Education Research." A 900-page summary of what we know about computing education. Cambridge University Press, 2019. ISBN: 978-1108721899.

[**Fink2013**] L. Dee Fink. "Creating Significant Learning Experiences: An Integrated Approach to Designing College Courses." A step-by-step guide to a systematic lesson design process. Jossey-Bass, 2013. ISBN: 1118124251.

[**Fisc2015**] Lars Fischer and Stefan Hanenberg. "An empirical investigation of the effects of type systems and code completion on API usability using TypeScript and JavaScript in MS Visual Studio." In: *11th Symposium on Dynamic Languages (DLS'15)*. Found that static typing improved programmer efficiency independently of code completion. ACM Press, 2015. DOI: 10.1145/2816707. 2816720.

[**Fisl2014**] Kathi Fisler. "The Recurring Rainfall Problem." In: *2014 International Computing Education Research Conference (ICER'14)*. Reports that students made fewer low-level errors when solving the Rainfall Problem in a functional language. Association for Computing Machinery (ACM), 2014. DOI: 10.1145/ 2632320.2632346.

[Fitz2008] Sue Fitzgerald, Gary Lewandowski, Renée McCauley, Laurie Murphy, Beth Simon, Lynda Thomas, and Carol Zander. "Debugging: Finding, Fixing and Flailing, a Multi-Institutional Study of Novice Debuggers." In: *Computer Science Education* 18.2 (June 2008). Reports that good undergraduate debuggers are good programmers but not necessarily vice versa, and that novices use tracing and testing rather than causal reasoning, pp. 93–116. DOI: 10.1080/08993400802114508.

[Foge2005] Karl Fogel. "Producing Open Source Software: How to Run a Successful Free Software Project." The definite guide to managing open source software development projects. O'Reilly Media, 2005. ISBN: 0596007590.

[Ford2016] Denae Ford, Justin Smith, Philip J. Guo, and Chris Parnin. "Paradise Unplugged: Identifying Barriers for Female Participation on Stack Overflow." In: *2016 International Symposium on Foundations of Software Engineering (FSE'16)*. Reports that lack of awareness of site features, feeling unqualified to answer questions, intimidating community size, discomfort interacting with or relying on strangers, and perception that they shouldn't be slacking were seen as significantly more problematic by female Stack Overflow contributors rather than male ones. Association for Computing Machinery (ACM), 2016. DOI: 10.1145/2950290.2950331.

[Fran2018] Pablo Frank-Bolton and Rahul Simha. "Docendo Discimus: Students Learn by Teaching Peers Through Video." In: *2018 Technical Symposium on Computer Science Education (SIGCSE'18)*. Reports that students who make short videos to teach concepts to their peers have a significant increase in their own learning compared to those who only study the material or view videos. Association for Computing Machinery (ACM), 2018. DOI: 10.1145/3159450.3159466.

[Free1972] Jo Freeman. "The Tyranny of Structurelessness." In: *The Second Wave* 2.1 (1972). Points out that every organization has a power structure: the only question is whether it's accountable or not.

[Free2014] S. Freeman, S. L. Eddy, M. McDonough, M. K. Smith, N. Okoroafor, H. Jordt, and M. P. Wenderoth. "Active learning increases student performance in science, engineering, and mathematics." In: *Proc. National Academy of Sciences* 111.23 (May 2014). Presents a meta-analysis of the benefits of active learning, pp. 8410–8415. DOI: 10.1073/pnas.1319030111.

[Frie2016] Marilyn Friend and Lynne Cook. "Interactions: Collaboration Skills for School Professionals." Eighth. A textbook on how teachers can work with other teachers. Pearson, 2016. ISBN: 0134168542.

[Galp2002] Vashti Galpin. "Women in Computing Around the World." In: *ACM SIGCSE Bulletin* 34.2 (June 2002). Looks at female participation in computing in 35 countries. DOI: 10.1145/543812.543839.

[Gauc2011] Danielle Gaucher, Justin Friesen, and Aaron C. Kay. "Evidence that Gendered Wording in Job Advertisements Exists and Sustains Gender Inequality." In: *Journal of Personality and Social Psychology* 101.1 (2011). Reports that gendered wording in job recruitment materials can maintain gender inequality in traditionally male-dominated occupations, pp. 109–128. DOI: 10. 1037/a0022530.

[Gawa2007] Atul Gawande. "The Checklist." In: *The New Yorker* (Dec. 2007). Describes the life-saving effects of simple checklists.

[Gawa2011] Atul Gawande. "Personal Best." In: *The New Yorker* (Oct. 2011). Describes how having a coach can improve practice in a variety of fields.

[Gick1987] Mary L. Gick and Keith J. Holyoak. "The Cognitive Basis of Knowledge Transfer." In: *Transfer of Learning: Contemporary Research and Applications*. Ed. by S. J. Cormier and J. D. Hagman. Finds that transference only comes with mastery. Elsevier, 1987, pp. 9–46. DOI: 10.1016/b978-0-12-188950-0.50008-4.

[Gorm2014] Cara Gormally, Mara Evans, and Peggy Brickman. "Feedback About Teaching in Higher Ed: Neglected Opportunities to Promote Change." In: *Cell Biology Education* 13.2 (June 2014). Summarizes best practices for providing instructional feedback, and recommends some specific strategies, pp. 187–199. DOI: 10.1187/cbe.13-12-0235.

[Gree2014] Elizabeth Green. "Building a Better Teacher: How Teaching Works (and How to Teach It to Everyone)." Explains why educational reforms in the past fifty years has mostly missed the mark, and what we should do instead. W. W. Norton & Company, 2014. ISBN: 0393351084.

[Grif2016] Jean M. Griffin. "Learning by Taking Apart." In: *2016 Conference on Information Technology Education (SIGITE'16)*. Reports that people learn to program more quickly by deconstructing code than by writing it. ACM Press, 2016. DOI: 10.1145/2978192.2978231.

[Grov2017] Shuchi Grover and Satabdi Basu. "Measuring Student Learning in Introductory Block-Based Programming." In: *2017 Technical Symposium on Computer Science Education (SIGCSE'17)*. Reports that middle-school children using blocks-based programming find loops, variables, and Boolean operators difficult to understand. Association for Computing Machinery (ACM), 2017. DOI: 10.1145/3017680.3017723.

[Gull2004] Ned Gulley. "In Praise of Tweaking." In: *interactions* 11.3 (May 2004). Describes an innovative collaborative coding contest, p. 18. DOI: 10.1145/986253.986264.

[Guo2013] Philip J. Guo. "Online Python Tutor." In: *2013 Technical Symposium on Computer Science Education (SIGCSE'13)*. Describes the design and use of a web-based execution visualization tool. Association for Computing Machinery (ACM), 2013. DOI: 10.1145/2445196.2445368.

[Guo2014] Philip J. Guo, Juho Kim, and Rob Rubin. "How Video Production Affects Student Engagement." In: *2014 Conference on Learning @ Scale (L@S'14)*. Measured learner engagement with MOOC videos and reports that short videos are more engaging than long ones and that talking heads are more engaging than tablet drawings. Association for Computing Machinery (ACM), 2014. DOI: 10.1145/2556325.2566239.

[Guzd2013] Mark Guzdial. "Exploring Hypotheses about Media Computation." In: *2013 International Computing Education Research Conference (ICER'13)*. A look back on ten years of media computation research. Association for Computing Machinery (ACM), 2013. DOI: 10.1145/2493394.2493397.

[Guzd2015a] Mark Guzdial. "Learner-Centered Design of Computing Education: Research on Computing for Everyone." Argues that we must design computing education for everyone, not just people who think they are going to become professional programmers. Morgan & Claypool Publishers, 2015. ISBN: 9781627053518.

[Guzd2015b] Mark Guzdial. *Top 10 Myths About Teaching Computer Science.* https://cacm.acm.org/blogs/blog-cacm/189498-top-10-myths-about-teaching-computer-science/fulltext. Ten things many people believe about teaching computing that simply aren't true. 2015.

[Guzd2016] Mark Guzdial. *Five Principles for Programming Languages for Learners.* https://cacm.acm.org/blogs/blog-cacm/203554-five-principles-for-programming-languages-for-learners/fulltext. Explains how to choose a programming language for people who are new to programming. 2016.

[Haar2017] Lassi Haaranen. "Programming as a Performance - Live-streaming and Its Implications for Computer Science Education." In: *2017 Conference on Innovation and Technology in Computer Science Education (ITiCSE'17)*. An early look at live streaming of coding as a teaching technique. Association for Computing Machinery (ACM), 2017. DOI: 10.1145/3059009.3059035.

[Hake1998] Richard R. Hake. "Interactive Engagement versus Traditional Methods: A Six-Thousand-Student Survey of Mechanics Test Data for Introductory Physics Courses." In: *American Journal of Physics* 66.1 (Jan. 1998). Reports the use of a concept inventory to measure the benefits of interactive engagement as a teaching technique, pp. 64–74. DOI: 10.1119/1.18809.

[Hamo2017] Sally Hamouda, Stephen H. Edwards, Hicham G. Elmongui, Jeremy V. Ernst, and Clifford A. Shaffer. "A Basic Recursion Concept Inventory." In: *Computer Science Education* 27.2 (Apr. 2017). Reports early work on developing a concept inventory for recursion, pp. 121–148. DOI: 10.1080/08993408.2017.1414728.

[Hank2011] Brian Hanks, Sue Fitzgerald, Renée McCauley, Laurie Murphy, and Carol Zander. "Pair Programming in Education: a Literature Review." In: *Computer Science Education* 21.2 (June 2011). Reports increased success rates and retention with pair programming, with some evidence that it is particularly beneficial for women, but finds that scheduling and partner compatibility can be problematic, pp. 135–173. DOI: 10.1080/08993408.2011.579808.

[Hann2009] Jo Erskine Hannay, Tore Dybå, Erik Arisholm, and Dag I. K. Sjøberg. "The Effectiveness of Pair Programming: A Meta-analysis." In: *Information and Software Technology* 51.7 (July 2009). A comprehensive meta-analysis of research on pair programming, pp. 1110–1122. DOI: 10.1016/j.infsof.2009.02. 001.

[Hann2010] Jo Erskine Hannay, Erik Arisholm, Harald Engvik, and Dag I. K. Sjøberg. "Effects of Personality on Pair Programming." In: *IEEE Transactions on Software Engineering* 36.1 (Jan. 2010). Reports weak correlation between the "Big Five" personality traits and performance in pair programming, pp. 61–80. DOI: 10.1109/tse.2009.41.

[Hans2015] John D. Hansen and Justin Reich. "Democratizing education? Examining access and usage patterns in massive open online courses." In: *Science* 350.6265 (Dec. 2015). Reports that MOOCs are mostly used by the affluent, pp. 1245–1248. DOI: 10.1126/science.aab3782.

[Harm2016] Kyle James Harms, Jason Chen, and Caitlin L. Kelleher. "Distractors in Parsons Problems Decrease Learning Efficiency for Young Novice Programmers." In: *2016 International Computing Education Research Conference (ICER'16)*. Shows that adding distractors to Parsons Problems does not improve learning outcomes but increases solution times. Association for Computing Machinery (ACM), 2016. DOI: 10.1145/2960310.2960314.

[Harr2018] Brian Harrington and Nick Cheng. "Tracing vs. Writing Code: Beyond the Learning Hierarchy." In: *2018 Technical Symposium on Computer Science Education (SIGCSE'18)*. Finds that the gap between being able to trace code and being able to write it has largely closed by CS2, and that students who still have a gap (in either direction) are likely to do poorly in the course. Association for Computing Machinery (ACM), 2018. DOI: 10.1145/3159450.3159530.

[Hazz2014] Orit Hazzan, Tami Lapidot, and Noa Ragonis. "Guide to Teaching Computer Science: An Activity-Based Approach." Second. A textbook for teaching computer science at the K-12 level with dozens of activities. Springer, 2014. ISBN: 9781447166290.

[Hend2015a] Charles Henderson, Renée Cole, Jeff Froyd, Debra Friedrichsen, Raina Khatri, and Courtney Stanford. "Designing Educational Innovations for Sustained Adoption." A detailed analysis of strategies for getting institutions in higher education to make changes. Increase the Impact, 2015. ISBN: 0996835210.

[Hend2015b] Charles Henderson, Renée Cole, Jeff Froyd, Debra Friedrichsen, Raina Khatri, and Courtney Stanford. *Designing Educational Innovations for Sustained Adoption (Executive Summary)*. http://www.increasetheimpact.com/resources.html. A short summary of key points from the authors' work on effecting change in higher education. 2015.

[Hend2017] Carl Hendrick and Robin Macpherson. "What Does This Look Like In The Classroom?: Bridging The Gap Between Research And Practice." A collection of responses by educational experts to questions asked by classroom teachers, with prefaces by the authors. John Catt Educational, 2017. ISBN: 9781911382379.

[**Henr2010**] Joseph Henrich, Steven J. Heine, and Ara Norenzayan. "The Weirdest People in the World?." In: *Behavioral and Brain Sciences* 33.2-3 (June 2010). Points out that the subjects of most published psychological studies are Western, educated, industrialized, rich, and democratic, pp. 61–83. DOI: 10.1017/s0140525x0999152x.

[**Hest1992**] David Hestenes, Malcolm Wells, and Gregg Swackhamer. "Force Concept Inventory." In: *The Physics Teacher* 30.3 (Mar. 1992). Describes the Force Concept Inventory's motivation, design, and impact, pp. 141–158. DOI: 10.1119/1.2343497.

[**Hick2018**] Marie Hicks. "Programmed Inequality: How Britain Discarded Women Technologists and Lost Its Edge in Computing." Describes how Britain lost its early dominance in computing by systematically discriminating against its most qualified workers: women. MIT Press, 2018. ISBN: 9780262535182.

[**Hofm2017**] Johannes Hofmeister, Janet Siegmund, and Daniel V. Holt. "Shorter Identifier Names Take Longer to Comprehend." In: *2017 Conference on Software Analysis, Evolution and Reengineering (SANER'17)*. Reports that using words for variable names makes comprehension faster than using abbreviations or single-letter names for variables. Institute of Electrical and Electronics Engineers (IEEE), Feb. 2017. DOI: 10.1109/saner.2017.7884623.

[**Holl1960**] Jack Hollingsworth. "Automatic Graders for Programming Classes." In: *Communications of the ACM* 3.10 (Oct. 1960). A brief note describing what may have been the world's first auto-grader, pp. 528–529. DOI: 10.1145/367415.367422.

[**Hpl2018**] National Academies of Sciences, Engineering, and Medicine. "How People Learn II: Learners, Contexts, and Cultures." A comprehensive survey of what we know about learning. National Academies Press, 2018. ISBN: 978-0309459648.

[**Hu2017**] Helen H. Hu, Cecily Heiner, Thomas Gagne, and Carl Lyman. "Building a Statewide Computer Science Teacher Pipeline." In: *2017 Technical Symposium on Computer Science Education (SIGCSE'17)*. Reports that a six-month program for high school teachers converting to teach CS quadruples the number of teachers without noticeable reduction of student outcomes and increases teachers' belief that anyone can program. Association for Computing Machinery (ACM), 2017. DOI: 10.1145/3017680.3017788.

[**Hust2012**] Therese Huston. "Teaching What You Don't Know." A pointed, funny, and very useful exploration of exactly what the title says. Harvard University Press, 2012. ISBN: 0674066170.

[**Ihan2010**] Petri Ihantola, Tuukka Ahoniemi, Ville Karavirta, and Otto Seppälä. "Review of Recent Systems for Automatic Assessment of Programming

Assignments." In: *2010 Koli Calling Conference on Computing Education Research (Koli'10)*. Reviews auto-grading tools of the time. Association for Computing Machinery (ACM), 2010. DOI: 10.1145/1930464.1930480.

[Ihan2011] Petri Ihantola and Ville Karavirta. "Two-dimensional Parson's Puzzles: The Concept, Tools, and First Observations." In: *Journal of Information Technology Education: Innovations in Practice* 10 (2011). Describes a 2D Parsons Problem tool and early experiences with it that confirm that experts solve outside-in rather than line-by-line, pp. 119–132. DOI: 10.28945/1394.

[Ihan2016] Petri Ihantola, Kelly Rivers, Miguel Ángel Rubio, Judy Sheard, Bronius Skupas, Jaime Spacco, Claudia Szabo, Daniel Toll, Arto Vihavainen, Alireza Ahadi, Matthew Butler, Jürgen Börstler, Stephen H. Edwards, Essi Isohanni, Ari Korhonen, and Andrew Petersen. "Educational Data Mining and Learning Analytics in Programming: Literature Review and Case Studies." In: *2016 Conference on Innovation and Technology in Computer Science Education (ITiCSE'16)*. A survey of methods used in mining and analyzing programming data. Association for Computing Machinery (ACM), 2016. DOI: 10.1145/2858796.2858798.

[Ijss2000] Wijnand A. IJsselsteijn, Huib de Ridder, Jonathan Freeman, and Steve E. Avons. "Presence: Concept, Determinants, and Measurement." In: *2000 Conference on Human Vision and Electronic Imaging*. Ed. by Bernice E. Rogowitz and Thrasyvoulos N. Pappas. Summarizes thinking of the time about real and virtual presence. SPIE, June 2000. DOI: 10.1117/12.387188.

[Irib2009] Alicia Iriberri and Gondy Leroy. "A Life-Cycle Perspective on Online Community Success." In: *ACM Computing Surveys* 41.2 (Feb. 2009). Reviews research on online communities organized according to a five-stage lifecycle model, pp. 1–29. DOI: 10.1145/1459352.1459356.

[Juss2005] Lee Jussim and Kent D. Harber. "Teacher Expectations and Self-Fulfilling Prophecies: Knowns and Unknowns, Resolved and Unresolved Controversies." In: *Personality and Social Psychology Review* 9.2 (May 2005). A survey of the effects of teacher expectations on student outcomes, pp. 131–155. DOI: 10.1207/s15327957pspr0902_3.

[Kaly2003] Slava Kalyuga, Paul Ayres, Paul Chandler, and John Sweller. "The Expertise Reversal Effect." In: *Educational Psychologist* 38.1 (Mar. 2003). Reports that instructional techniques that work well with inexperienced learners lose their effectiveness or have negative consequences when used with more experienced learners, pp. 23–31. DOI: 10.1207/s15326985ep3801_4.

[Kaly2015] Slava Kalyuga and Anne-Marie Singh. "Rethinking the Boundaries of Cognitive Load Theory in Complex Learning." In: *Educational Psychology Review* 28.4 (Dec. 2015). Argues that cognitive load theory is basically micromanagement within a broader pedagogical context, pp. 831–852. DOI: 10.1007/s10648-015-9352-0.

[Kang2016] Sean H. K. Kang. "Spaced Repetition Promotes Efficient and Effective Learning." In: *Policy Insights from the Behavioral and Brain Sciences* 3.1 (Jan. 2016). Summarizes research on spaced repetition and what it means for classroom teaching, pp. 12–19. DOI: 10.1177/2372732215624708.

[Kapu2016] Manu Kapur. "Examining Productive Failure, Productive Success, Unproductive Failure, and Unproductive Success in Learning." In: *Educational Psychologist* 51.2 (Apr. 2016). Looks at productive failure as an alternative to inquiry-based learning and approaches based on cognitive load theory, pp. 289–299. DOI: 10.1080/00461520.2016.1155457.

[Karp2008] Jeffrey D. Karpicke and Henry L. Roediger. "The Critical Importance of Retrieval for Learning." In: *Science* 319.5865 (Feb. 2008). Reports that repeated testing improves recall of word lists from 35% to 80%, even when learners can still access the material but are not tested on it, pp. 966–968. DOI: 10.1126/science.1152408.

[Kauf2000] Deborah B. Kaufman and Richard M. Felder. "Accounting for Individual Effort in Cooperative Learning Teams." In: *Journal of Engineering Education* 89.2 (2000). Reports that self-rating and peer ratings in undergraduate courses agree, that collusion isn't significant, that students don't inflate their self-ratings, and that ratings are not biased by gender or race.

[Keme2009] Chris F. Kemerer and Mark C. Paulk. "The Impact of Design and Code Reviews on Software Quality: An Empirical Study Based on PSP Data." In: *IEEE Transactions on Software Engineering* 35.4 (July 2009). Uses individual data to explore the effectiveness of code review, pp. 534–550. DOI: 10.1109/tse.2009.27.

[Kepp2008] Jeroen Keppens and David Hay. "Concept Map Assessment for Teaching Computer Programming." In: *Computer Science Education* 18.1 (Mar. 2008). A short review of ways concept mapping can be used in CS education, pp. 31–42. DOI: 10.1080/08993400701864880.

[Kern1978] Brian W. Kernighan and P. J. Plauger. "The Elements of Programming Style." Second. An early and influential description of the Unix programming philosophy. McGraw-Hill, 1978. ISBN: 0070342075.

[Kern1983] Brian W. Kernighan and Rob Pike. "The Unix Programming Environment." An influential early description of Unix. Prentice-Hall, 1983. ISBN: 013937681X.

[Kern1988] Brian W. Kernighan and Dennis M. Ritchie. "The C Programming Language." Second. The book that made C a popular programming language. Prentice-Hall, 1988. ISBN: 0131103628.

[Kern1999] Brian W. Kernighan and Rob Pike. "The Practice of Programming." A programming style manual written by two of the creators of modern computing. Addison-Wesley, 1999. ISBN: 9788177582482.

[Keun2016a] Hieke Keuning, Johan Jeuring, and Bastiaan Heeren. "Towards a Systematic Review of Automated Feedback Generation for Programming Exercises." In: *2016 Conference on Innovation and Technology in Computer Science Education (ITiCSE'16)*. Reports that auto-grading tools often do not give feedback on what to do next, and that teachers cannot easily adapt most of the tools to their needs. Association for Computing Machinery (ACM), 2016. DOI: 10.1145/2899415.2899422.

[Keun2016b] Hieke Keuning, Johan Jeuring, and Bastiaan Heeren. *Towards a Systematic Review of Automated Feedback Generation for Programming Exercises - Extended Version*. Technical Report UU-CS-2016-001, Utrecht University. An extended look at feedback messages from auto-grading tools. 2016.

[Kim2017] Ada S. Kim and Andrew J. Ko. "A Pedagogical Analysis of Online Coding Tutorials." In: *2017 Technical Symposium on Computer Science Education (SIGCSE'17)*. Reports that online coding tutorials largely teach similar content, organize content bottom-up, and provide goal-directed practices with immediate feedback, but are not tailored to learners' prior coding knowledge and usually don't tell learners how to transfer and apply knowledge. Association for Computing Machinery (ACM), 2017. DOI: 10.1145/3017680.3017728.

[King1993] Alison King. "From Sage on the Stage to Guide on the Side." In: *College Teaching* 41.1 (Jan. 1993). An early proposal to flip the classroom, pp. 30–35. DOI: 10.1080/87567555.1993.9926781.

[Kirk1994] Donald L. Kirkpatrick. "Evaluating Training Programs: The Four Levels." Defines a widely used four-level model for evaluating training. Berrett-Koehle, 1994. ISBN: 1881052494.

[Kirs2006] Paul A. Kirschner, John Sweller, and Richard E. Clark. "Why Minimal Guidance During Instruction does not Work: An Analysis of the Failure of Constructivist, Discovery, Problem-Based, Experiential, and Inquiry-Based Teaching." In: *Educational Psychologist* 41.2 (June 2006). Argues that inquiry-based learning is less effective for novices than guided instruction, pp. 75–86. DOI: 10.1207/s15326985ep4102_1.

[Kirs2013] Paul A. Kirschner and Jeroen J. G. van Merriënboer. "Do Learners Really Know Best? Urban Legends in Education." In: *Educational Psychologist* 48.3 (July 2013). Argues that three learning myths—digital natives, learning styles, and self-educators—all reflect the mistaken belief that learners know what is best for them, and cautions that we may be in a downward spiral in which every attempt by education researchers to rebut these myths confirms their opponents' belief that learning science is pseudo-science, pp. 169–183. DOI: 10.1080/00461520.2013.804395.

[Kirs2018] Paul A. Kirschner, John Sweller, Femke Kirschner, and Jimmy Zambrano R. "From Cognitive Load Theory to Collaborative Cognitive Load Theory." In: *International Journal of Computer-Supported Collaborative Learning* (Apr. 2018). Extends cognitive load theory to include collaborative aspects of learning. DOI: 10.1007/s11412-018-9277-y.

[**Koed2015**] Kenneth R. Koedinger, Jihee Kim, Julianna Zhuxin Jia, Elizabeth A. McLaughlin, and Norman L. Bier. "Learning is Not a Spectator Sport: Doing is Better than Watching for Learning from a MOOC." In: *2015 Conference on Learning @ Scale (L@S'15)*. Measures the benefits of doing rather than watching. Association for Computing Machinery (ACM), 2015. DOI: 10.1145/2724660.2724681.

[**Koeh2013**] Matthew J. Koehler, Punya Mishra, and William Cain. "What is Technological Pedagogical Content Knowledge (TPACK)?." In: *Journal of Education* 193.3 (2013). Refines the discussion of PCK by adding technology, and sketches strategies for building understanding of how to use it, pp. 13–19. DOI: 10.1177/002205741319300303.

[**Kohn2017**] Tobias Kohn. "Variable Evaluation: An Exploration of Novice Programmers' Understanding and Common Misconceptions." In: *2017 Technical Symposium on Computer Science Education (SIGCSE'17)*. Reports that students often believe in delayed evaluation or that entire equations are stored in variables. Association for Computing Machinery (ACM), 2017. DOI: 10.1145/3017680.3017724.

[**Koll2015**] Michael Kölling. "Lessons From the Design of Three Educational Programming Environments." In: *International Journal of People-Oriented Programming* 4.1 (Jan. 2015). Compares three generations of programming environments intended for novice use, pp. 5–32. DOI: 10.4018/ijpop.2015010102.

[**Krau2016**] Robert E. Kraut and Paul Resnick. "Building Successful Online Communities: Evidence-Based Social Design." Sums up what we actually know about making thriving online communities and why we believe it's true. MIT Press, 2016. ISBN: 0262528916.

[**Krug1999**] Justin Kruger and David Dunning. "Unskilled and Unaware of it: How Difficulties in Recognizing One's Own Incompetence Lead to Inflated Self-Assessments." In: *Journal of Personality and Social Psychology* 77.6 (1999). The original report on the Dunning-Kruger effect: the less people know, the less accurate their estimate of their knowledge, pp. 1121–1134. DOI: 10.1037/0022-3514.77.6.1121.

[**Kuch2011**] Marc J. Kuchner. "Marketing for Scientists: How to Shine in Tough Times." A short, readable guide to making people aware of, and care about, your work. Island Press, 2011. ISBN: 1597269948.

[**Kuit2004**] Marja Kuittinen and Jorma Sajaniemi. "Teaching Roles of Variables in Elementary Programming Courses." In: *ACM SIGCSE Bulletin* 36.3 (Sept. 2004). Presents a few patterns used in novice programming and the pedagogical value of teaching them, p. 57. DOI: 10.1145/1026487.1008014.

[**Kulk2013**] Chinmay Kulkarni, Koh Pang Wei, Huy Le, Daniel Chia, Kathryn Papadopoulos, Justin Cheng, Daphne Koller, and Scott R. Klemmer. "Peer and Self Assessment in Massive Online Classes." In: *ACM Transactions on*

Computer-Human Interaction 20.6 (Dec. 2013). Shows that peer grading can be as effective at scale as expert grading, pp. 1–31. DOI: 10.1145/2505057.

[**Laba2008**] David F. Labaree. "The Winning Ways of a Losing Strategy: Educationalizing Social Problems in the United States." In: *Educational Theory* 58.4 (Nov. 2008). Explores why the United States keeps pushing the solution of social problems onto educational institutions, and why that continues not to work, pp. 447–460. DOI: 10.1111/j.1741-5446.2008.00299.x.

[**Lach2018**] Michael Lachney. "Computational Communities: African-American Cultural Capital in Computer Science Education." In: *Computer Science Education* (Feb. 2018). Explores use of community representation and computational integration to bridge computing and African-American cultural capital in CS education, pp. 1–22. DOI: 10.1080/08993408.2018.1429062.

[**Lake2018**] George Lakey. "How We Win: A Guide to Nonviolent Direct Action Campaigning." A short experience-based guide to effective campaigning. Melville House, 2018. ISBN: 978-1612197531.

[**Lang2013**] James M. Lang. "Cheating Lessons: Learning from Academic Dishonesty." Explores why students cheat, and how courses often give them incentives to do so. Harvard University Press, 2013. ISBN: 0674724631.

[**Lang2016**] James M. Lang. "Small Teaching: Everyday Lessons from the Science of Learning." Presents a selection of accessible evidence-based practices that teachers can adopt when they have little time and few resources. Jossey-Bass, 2016. ISBN: 9781118944493.

[**Lazo1993**] Ard W. Lazonder and Hans van der Meij. "The Minimal Manual: Is Less Really More?." In: *International Journal of Man-Machine Studies* 39.5 (Nov. 1993). Reports that the minimal manual approach to instruction outperforms traditional approaches regardless of prior experience with computers, pp. 729–752. DOI: 10.1006/imms.1993.1081.

[**Leak2017**] Mackenzie Leake and Colleen M. Lewis. "Recommendations for Designing CS Resource Sharing Sites for All Teachers." In: *2017 Technical Symposium on Computer Science Education (SIGCSE'17)*. Explores why CS teachers don't use resource sharing sites and recommends ways to make them more appealing. Association for Computing Machinery (ACM), 2017. DOI: 10.1145/3017680.3017780.

[**Lee2013**] Cynthia Bailey Lee. "Experience Report: CS1 in MATLAB for Non-Majors, with Media Computation and Peer Instruction." In: *2013 Technical Symposium on Computer Science Education (SIGCSE'13)*. Describes an adaptation of media computation to a first-year MATLAB course. Association for Computing Machinery (ACM), 2013. DOI: 10.1145/2445196.2445214.

[**Lee2017**] Cynthia Bailey Lee. *What Can I Do Today to Create a More Inclusive Community in CS?* http://bit.ly/2oynmSH. A practical checklist of things instructors can do to make their computing classes more inclusive. 2017.

[Lemo2014] Doug Lemov. "Teach Like a Champion 2.0: 62 Techniques that Put Students on the Path to College." Presents 62 classroom techniques drawn from intensive study of thousands of hours of video of good teachers in action. Jossey-Bass, 2014. ISBN: 1118901851.

[Lewi2015] Colleen M. Lewis and Niral Shah. "How Equity and Inequity Can Emerge in Pair Programming." In: *2015 International Computing Education Research Conference (ICER'15)*. Reports a study of pair programming in a middle-grade classroom in which less equitable pairs were ones that sought to complete the task quickly. Association for Computing Machinery (ACM), 2015. DOI: 10.1145/2787622.2787716.

[List2004] Raymond Lister, Otto Seppälä, Beth Simon, Lynda Thomas, Elizabeth S. Adams, Sue Fitzgerald, William Fone, John Hamer, Morten Lindholm, Robert McCartney, Jan Erik Moström, and Kate Sanders. "A Multi-National Study of Reading and Tracing Skills in Novice Programmers." In: *2004 Conference on Innovation and Technology in Computer Science Education (ITiCSE'04)*. Reports that students are weak at both predicting the outcome of executing a short piece of code and at selecting the correct completion for short pieces of code. Association for Computing Machinery (ACM), 2004. DOI: 10.1145/1044550.1041673.

[List2009] Raymond Lister, Colin Fidge, and Donna Teague. "Further Evidence of a Relationship Between Explaining, Tracing and Writing Skills in Introductory Programming." In: *ACM SIGCSE Bulletin* 41.3 (Aug. 2009). Replicates earlier studies showing that students who cannot trace code usually cannot explain code and that students who tend to perform reasonably well at code writing tasks have also usually acquired the ability to both trace code and explain code, p. 161. DOI: 10.1145/1595496.1562930.

[Litt2004] Dennis Littky. "The Big Picture: Education Is Everyone's Business." Essays on the purpose of education and how to make schools better. Association for Supervision & Curriculum Development (ASCD), 2004. ISBN: 0871209713.

[Luxt2009] Andrew Luxton-Reilly. "A Systematic Review of Tools That Support Peer Assessment." In: *Computer Science Education* 19.4 (Dec. 2009). Surveys peer assessment tools that may be of use in computing education, pp. 209–232. DOI: 10.1080/08993400903384844.

[Luxt2017] Andrew Luxton-Reilly, Jacqueline Whalley, Brett A. Becker, Yingjun Cao, Roger McDermott, Claudio Mirolo, Andreas Mühling, Andrew Petersen, Kate Sanders, and Simon. "Developing Assessments to Determine Mastery of Programming Fundamentals." In: *2017 Conference on Innovation and Technology in Computer Science Education (ITiCSE'17)*. Synthesizes work from many previous works to determine what CS instructors are actually teaching, how those things depend on each other, and how they might be assessed. Association for Computing Machinery (ACM), 2017. DOI: 10.1145/3174781.3174784.

[**Macn2014**] Brooke N. Macnamara, David Z. Hambrick, and Frederick L. Oswald. "Deliberate Practice and Performance in Music, Games, Sports, Education, and Professions: A Meta-Analysis." In: *Psychological Science* 25.8 (July 2014). A meta-study of the effectiveness of deliberate practice, pp. 1608–1618. DOI: 10.1177/0956797614535810.

[**Magu2018**] Phil Maguire, Rebecca Maguire, and Robert Kelly. "Using Automatic Machine Assessment to Teach Computer Programming." In: *Computer Science Education* (Feb. 2018). Reports that weekly machine-evaluated tests are a better predictor of exam scores than labs (but that students didn't like the system), pp. 1–18. DOI: 10.1080/08993408.2018.1435113.

[**Majo2015**] Claire Howell Major, Michael S. Harris, and Tod Zakrajsek. "Teaching for Learning: 101 Intentionally Designed Educational Activities to Put Students on the Path to Success." Catalogs a hundred different kinds of exercises to do with students. Routledge, 2015. ISBN: 0415699363.

[**Malo2010**] John Maloney, Mitchel Resnick, Natalie Rusk, Brian Silverman, and Evelyn Eastmond. "The Scratch Programming Language and Environment." In: *ACM Transactions on Computing Education* 10.4 (Nov. 2010). Summarizes the design of the first generation of Scratch, pp. 1–15. DOI: 10.1145/1868358. 1868363.

[**Mann2015**] Mary Lynn Manns and Linda Rising. "Fearless Change: Patterns for Introducing New Ideas." A catalog of patterns for making change happen in large organizations. Addison-Wesley, 2015. ISBN: 9780201741575.

[**Marc2011**] Guillaume Marceau, Kathi Fisler, and Shriram Krishnamurthi. "Measuring the Effectiveness of Error Messages Designed for Novice Programmers." In: *2011 Technical Symposium on Computer Science Education (SIGCSE'11)*. Looks at edit-level responses to error messages, and introduces a useful rubric for classifying user responses to errors. Association for Computing Machinery (ACM), 2011. DOI: 10.1145/1953163.1953308.

[**Marg2003**] Jane Margolis and Allan Fisher. "Unlocking the Clubhouse: Women in Computing." A groundbreaking report on the gender imbalance in computing, and the steps Carnegie Mellon took to address the problem. MIT Press, 2003. ISBN: 0262632691.

[**Marg2010**] Jane Margolis, Rachel Estrella, Joanna Goode, Jennifer Jellison Holme, and Kim Nao. "Stuck in the Shallow End: Education, Race, and Computing." Dissects the school structures and belief systems that lead to under-representation of African American and Latinx students in computing. MIT Press, 2010. ISBN: 0262514044.

[**Marg2012**] Lauren E. Margulieux, Mark Guzdial, and Richard Catrambone. "Subgoal-labeled Instructional Material Improves Performance and Transfer in Learning to Develop Mobile Applications." In: *2012 International Computing Education Research Conference (ICER'12)*. Reports that labeled subgoals improve outcomes and transference when learning about mobile app development. ACM Press, 2012, pp. 71–78. DOI: 10.1145/2361276.2361291.

[**Marg2015**] Anoush Margaryan, Manuela Bianco, and Allison Littlejohn. "Instructional Quality of Massive Open Online Courses (MOOCs)." In: *Computers & Education* 80 (Jan. 2015). Reports that instructional design quality in MOOCs poor, but that the organization and presentation of material is good, pp. 77–83. DOI: 10.1016/j.compedu.2014.08.005.

[**Marg2016**] Lauren E. Margulieux, Richard Catrambone, and Mark Guzdial. "Employing Subgoals in Computer Programming Education." In: *Computer Science Education* 26.1 (Jan. 2016). Reports that labeled subgoals improve learning outcomes in introductory computing courses, pp. 44–67. DOI: 10. 1080/08993408.2016.1144429.

[**Mark2018**] Rebecca A. Markovits and Yana Weinstein. "Can Cognitive Processes Help Explain the Success of Instructional Techniques Recommended by Behavior Analysts?." In: *NPJ Science of Learning* 3.1 (Jan. 2018). Points out that behavioralists and cognitive psychologists differ in approach, but wind up making very similar recommendations about how to teach, and gives two specific examples. DOI: 10.1038/s41539-017-0018-1.

[**Mars2002**] Herbert W. Marsh and John Hattie. "The Relation Between Research Productivity and Teaching Effectiveness: Complementary, Antagonistic, or Independent Constructs?." In: *Journal of Higher Education* 73.5 (2002). One study of many showing there is zero correlation between research ability and teaching effectiveness, pp. 603–641. DOI: 10.1353/jhe.2002.0047.

[**Masa2018**] Susana Masapanta-Carrión and J. Ángel Velázquez-Iturbide. "A Systematic Review of the Use of Bloom's Taxonomy in Computer Science Education." In: *2018 Technical Symposium on Computer Science Education (SIGCSE'18)*. Reports that even experienced educators have trouble agreeing on the correct classification for a question or idea using Bloom's Taxonomy. Association for Computing Machinery (ACM), 2018. DOI: 10.1145/3159450. 3159491.

[**Maso2016**] Raina Mason, Carolyn Seton, and Graham Cooper. "Applying Cognitive Load Theory to the Redesign of a Conventional Database Systems Course." In: *Computer Science Education* 26.1 (Jan. 2016). Reports how redesigning a database course using cognitive load theory reduced exam failure rate while increasing student satisfaction, pp. 68–87. DOI: 10.1080/08993408. 2016.1160597.

[**Matt2019**] Eric Matthes. *Python Flash Cards: Syntax, Concepts, and Examples.* Handy flashcards summarizing the core of Python 3. 2019.

[**Maye2003**] Richard E. Mayer and Roxana Moreno. "Nine Ways to Reduce Cognitive Load in Multimedia Learning." In: *Educational Psychologist* 38.1 (Mar. 2003). Shows how research into how we absorb and process information can be applied to the design of instructional materials, pp. 43–52. DOI: 10.1207/s15326985ep3801_6.

[Maye2004] Richard E. Mayer. "Teaching of Subject Matter." In: *Annual Review of Psychology* 55.1 (Feb. 2004). An overview of how and why teaching and learning are subject-specific, pp. 715–744. DOI: 10.1146/annurev.psych.55.082602.133124.

[Maye2009] Richard E. Mayer. "Multimedia Learning." Second. Presents a cognitive theory of multimedia learning. Cambridge University Press, 2009. ISBN: 9780521735353.

[Mazu1996] Eric Mazur. "Peer Instruction: A User's Manual." A guide to implementing peer instruction. Prentice-Hall, 1996.

[McCa2008] Renée McCauley, Sue Fitzgerald, Gary Lewandowski, Laurie Murphy, Beth Simon, Lynda Thomas, and Carol Zander. "Debugging: A Review of the Literature from an Educational Perspective." In: *Computer Science Education* 18.2 (June 2008). Summarizes research about why bugs occur, why types there are, how people debug, and whether we can teach debugging skills, pp. 67–92. DOI: 10.1080/08993400802114581.

[McCr2001] Michael McCracken, Tadeusz Wilusz, Vicki Almstrum, Danny Diaz, Mark Guzdial, Dianne Hagan, Yifat Ben-David Kolikant, Cary Laxer, Lynda Thomas, and Ian Utting. "A Multi-National, Multi-Institutional Study of Assessment of Programming Skills of First-Year CS Students." In: *2001 Conference on Innovation and Technology in Computer Science Education (ITiCSE'01)*. Reports that most students still struggle to solve even basic programming problems at the end of their introductory course. Association for Computing Machinery (ACM), 2001. DOI: 10.1145/572133.572137.

[McDo2006] Charlie McDowell, Linda Werner, Heather E. Bullock, and Julian Fernald. "Pair Programming Improves Student Retention, Confidence, and Program Quality." In: *Communications of the ACM* 49.8 (Aug. 2006). A summary of research showing that pair programming improves retention and confidence, pp. 90–95. DOI: 10.1145/1145287.1145293.

[McGu2015] Saundra Yancey McGuire. "Teach Students How to Learn: Strategies You Can Incorporate Into Any Course to Improve Student Metacognition, Study Skills, and Motivation." Explains how metacognitive strategies can improve learning. Stylus Publishing, 2015. ISBN: 162036316X.

[McMi2017] Tressie McMillan Cottom. "Lower Ed: The Troubling Rise of For-Profit Colleges in the New Economy." Lays bare the dynamics of the growing educational industry to show how it leads to greater inequality rather than less. The New Press, 2017. ISBN: 1620970600.

[McTi2013] Jay McTighe and Grant Wiggins. *Understanding by Design Framework*. http://www.ascd.org/ASCD/pdf/siteASCD/publications/UbD_WhitePaper0312.pdf. Summarizes the backward instructional design process. 2013.

[Metc2016] Janet Metcalfe. "Learning from Errors." In: *Annual Review of Psychology* 68.1 (Jan. 2016). Summarizes work on the hypercorrection effect in learning, pp. 465–489. DOI: 10.1146/annurev-psych-010416-044022.

[Meys2018] Mark Meysenburg, Tessa Durham Brooks, Raychelle Burks, Erin Doyle, and Timothy Frey. "DIVAS: Outreach to the Natural Sciences Through Image Processing." In: *2018 Technical Symposium on Computer Science Education (SIGCSE'18)*. Describes early results from a programming course for science undergrads built around image processing. Association for Computing Machinery (ACM), 2018. DOI: 10.1145/3159450.3159537.

[Midw2010] Midwest Academy. "Organizing for Social Change: Midwest Academy Manual for Activists." Fourth. A training manual for people building progressive social movements. The Forum Press, 2010. ISBN: 0984275215.

[Mill1956] George A. Miller. "The Magical Number Seven, Plus or Minus Two: Some Limits on Our Capacity for Processing Information." In: *Psychological Review* 63.2 (1956). The original paper on the limited size of short-term memory, pp. 81–97. DOI: 10.1037/h0043158.

[Mill2013] Kelly Miller, Nathaniel Lasry, Kelvin Chu, and Eric Mazur. "Role of Physics Lecture Demonstrations in Conceptual Learning." In: *Physical Review Special Topics - Physics Education Research* 9.2 (Sept. 2013). Reports a detailed study of what students learn during demonstrations and why. DOI: 10.1103/physrevstper.9.020113.

[Mill2015] David I. Miller and Jonathan Wai. "The Bachelor's to Ph.D. STEM Pipeline No Longer Leaks More Women Than Men: a 30-year Analysis." In: *Frontiers in Psychology* 6 (Feb. 2015). Shows that the "leaky pipeline" metaphor stopped being accurate some time in the 1990s. DOI: 10.3389/fpsyg.2015.00037.

[Mill2016a] Michelle D. Miller. "Minds Online: Teaching Effectively with Technology." Describes ways that insights from neuroscience can be used to improve online teaching. Harvard University Press, 2016. ISBN: 0674660021.

[Mill2016b] Craig S. Miller and Amber Settle. "Some Trouble with Transparency: An Analysis of Student Errors with Object-Oriented Python." In: *2016 International Computing Education Research Conference (ICER'16)*. Reports that students have difficulty with self in Python. Association for Computing Machinery (ACM), 2016. DOI: 10.1145/2960310.2960327.

[Milt2018] Kate M. Miltner. "Girls Who Coded: Gender in Twentieth Century U.K. and U.S. Computing." In: *Science, Technology, & Human Values* (May 2018). A review of three books about how women were systematically pushed out of computing. DOI: 10.1177/0162243918770287.

[Mina1986] Anne Minahan. "Martha's Rules." In: *Affilia* 1.2 (June 1986). Describes a lightweight set of rules for consensus-based decision making, pp. 53–56. DOI: 10.1177/088610998600100206.

[Miya2018] Toshiya Miyatsu, Khuyen Nguyen, and Mark A. McDaniel. "Five Popular Study Strategies: Their Pitfalls and Optimal Implementations." In: *Perspectives on Psychological Science* 13.3 (May 2018). Explains how learners misuse common study strategies and what they should do instead, pp. 390–407. DOI: 10.1177/1745691617710510.

[Mlad2017] Monika Mladenović, Ivica Boljat, and Žana Žanko. "Comparing Loops Misconceptions in Block-Based and Text-Based Programming Languages at the K-12 Level." In: *Education and Information Technologies* (Nov. 2017). Reports that K-12 students have fewer misconceptions about loops using Scratch than using Logo or Python, and fewer misconceptions about nested loops with Logo than with Python. DOI: 10.1007/s10639-017-9673-3.

[More2019] Kayla Morehead, John Dunlosky, and Katherine A. Rawson. "How Much Mightier Is the Pen than the Keyboard for Note-Taking? A Replication and Extension of Mueller and Oppenheimer (2014)." In: *Educational Psychology Review* (Feb. 2019). Reports a failure to replicate an earlier study comparing note-taking by hand and with computers. DOI: 10.1007/s10648-019-09468-2.

[Morr2016] Briana B. Morrison, Lauren E. Margulieux, Barbara J. Ericson, and Mark Guzdial. "Subgoals Help Students Solve Parsons Problems." In: *2016 Technical Symposium on Computer Science Education (SIGCSE'16)*. Reports that students using labeled subgoals solve Parsons Problems better than students without labeled subgoals. Association for Computing Machinery (ACM), 2016. DOI: 10.1145/2839509.2844617.

[Muel2014] Pam A. Mueller and Daniel M. Oppenheimer. "The Pen is Mightier than the Keyboard." In: *Psychological Science* 25.6 (Apr. 2014). Presents evidence that taking notes by hand is more effective than taking notes on a laptop, pp. 1159–1168. DOI: 10.1177/0956797614524581.

[Mull2007a] Derek A. Muller, James Bewes, Manjula D. Sharma, and Peter Reimann. "Saying the Wrong Thing: Improving Learning with Multimedia by Including Misconceptions." In: *Journal of Computer Assisted Learning* 24.2 (July 2007). Reports that including explicit discussion of misconceptions significantly improves learning outcomes: students with low prior knowledge benefit most and students with more prior knowledge are not disadvantaged, pp. 144–155. DOI: 10.1111/j.1365-2729.2007.00248.x.

[Mull2007b] Orna Muller, David Ginat, and Bruria Haberman. "Pattern-Oriented Instruction and Its Influence on Problem Decomposition and Solution Construction." In: *2007 Technical Symposium on Computer Science Education (SIGCSE'07)*. Reports that explicitly teaching solution patterns improves learning outcomes. Association for Computing Machinery (ACM), 2007. DOI: 10.1145/1268784.1268830.

[Murp2008] Laurie Murphy, Gary Lewandowski, Renée McCauley, Beth Simon, Lynda Thomas, and Carol Zander. "Debugging: The Good, the Bad, and the Quirky - A Qualitative Analysis of Novices' Strategies." In: *ACM SIGCSE*

Bulletin 40.1 (Feb. 2008). Reports that many CS1 students use good debugging strategies, but many others don't, and students often don't recognize when they are stuck, p. 163. DOI: 10.1145/1352322.1352191.

[**Nara2018**] Sathya Narayanan, Kathryn Cunningham, Sonia Arteaga, William J. Welch, Leslie Maxwell, Zechariah Chawinga, and Bude Su. "Upward Mobility for Underrepresented Students." In: *2018 Technical Symposium on Computer Science Education (SIGCSE'18).* Describes an intensive 3-year bachelor's program based on tight-knit cohorts and administrative support that tripled graduation rates. Association for Computing Machinery (ACM), 2018. DOI: 10.1145/3159450.3159551.

[**Nath2003**] Mitchell J. Nathan and Anthony Petrosino. "Expert Blind Spot Among Preservice Teachers." In: *American Educational Research Journal* 40.4 (Jan. 2003). Early work on expert blind spot, pp. 905–928. DOI: 10.3102/00028312040004905.

[**Nils2017**] Linda B. Nilson and Ludwika A. Goodson. "Online Teaching at Its Best: Merging Instructional Design with Teaching and Learning Research." A guide for college instructors that focuses on online teaching. Jossey-Bass, 2017. ISBN: 1119242290.

[**Nord2017**] Emily Nordmann, Colin Calder, Paul Bishop, Amy Irwin, and Darren Comber. *Turn Up, Tune In, Don'T Drop Out: The Relationship Between Lecture Attendance, Use of Lecture Recordings, and Achievement at Different Levels of Study.* https://psyarxiv.com/fd3yj. Reports on the pros and cons of recording lectures. 2017. DOI: 10.17605/OSF.IO/FD3YJ.

[**Nutb2016**] Stephen Nutbrown and Colin Higgins. "Static Analysis of Programming Exercises: Fairness, Usefulness and a Method for Application." In: *Computer Science Education* 26.2-3 (May 2016). Describes ways auto-grader rules were modified and grades weighted to improve correlation between automatic feedback and manual grades, pp. 104–128. DOI: 10.1080/08993408.2016.1179865.

[**Nuth2007**] Graham Nuthall. "The Hidden Lives of Learners." Summarizes a lifetime of work looking at what students actually do in classrooms and how they actually learn. NZCER Press, 2007. ISBN: 1877398241.

[**Ojos2015**] Bobby Ojose. "Common Misconceptions in Mathematics: Strategies to Correct Them." A catalog of K-12 misconceptions in mathematics and what to do about them. UPA, 2015. ISBN: 0761858857.

[**Ornd2015**] Harold N. Orndorff III. "Collaborative Note-Taking: The Impact of Cloud Computing on Classroom Performance." In: *International Journal of Teaching and Learning in Higher Education* 27.3 (2015). Reports that taking notes together online is more effective than solo note-taking, pp. 340–351.

[**Ostr2015**] Elinor Ostrom. "Governing the Commons: The Evolution of Institutions for Collective Action." A masterful description and analysis of cooperative governance. Cambridge University Press, 2015. ISBN: 978-1107569782.

[Pape1993] Seymour A. Papert. "Mindstorms: Children, Computers, and Powerful Ideas." Second. The foundational text on how computers can underpin a new kind of education. Basic Books, 1993. ISBN: 0465046746.

[Pare2008] Dwayne E. Paré and Steve Joordens. "Peering Into Large Lectures: Examining Peer and Expert Mark Agreement Using peerScholar, an Online Peer Assessment Tool." In: *Journal of Computer Assisted Learning* 24.6 (Oct. 2008). Shows that peer grading by small groups can be as effective as expert grading once accountability features are introduced, pp. 526–540. DOI: 10.1111/j.1365-2729.2008.00290.x.

[Park2015] Thomas H. Park, Brian Dorn, and Andrea Forte. "An Analysis of HTML and CSS Syntax Errors in a Web Development Course." In: *ACM Transactions on Computing Education* 15.1 (Mar. 2015). Describes the errors students make in an introductory course on HTML and CSS, pp. 1–21. DOI: 10.1145/2700514.

[Park2016] Miranda C. Parker, Mark Guzdial, and Shelly Engleman. "Replication, Validation, and Use of a Language Independent CS1 Knowledge Assessment." In: *2016 International Computing Education Research Conference (ICER'16)*. Describes construction and replication of a second concept inventory for basic computing knowledge. Association for Computing Machinery (ACM), 2016. DOI: 10.1145/2960310.2960316.

[Parn1986] David Lorge Parnas and Paul C. Clements. "A Rational Design Process: How and Why to Fake It." In: *IEEE Transactions on Software Engineering* SE-12.2 (Feb. 1986). Argues that using a rational design process is less important than looking as though you had, pp. 251–257. DOI: 10.1109/tse.1986.6312940.

[Parn2017] Chris Parnin, Janet Siegmund, and Norman Peitek. "On the Nature of Programmer Expertise." In: *Psychology of Programming Interest Group Workshop 2017*. An annotated exploration of what "expertise" means in programming. 2017.

[Pars2006] Dale Parsons and Patricia Haden. "Parson's Programming Puzzles: A Fun and Effective Learning Tool for First Programming Courses." In: *2006 Australasian Conference on Computing Education (ACE'06)*. The first description of Parson's Problems. Australian Computer Society, 2006, pp. 157–163.

[Part2011] Anu Partanen. *What Americans Keep Ignoring About Finland's School Success*. https://www.theatlantic.com/national/archive/2011/12/what-americans-keep-ignoring-about-finlands-school-success/250564/. Explains that other countries struggle to replicate the success of Finland's schools because they're unwilling to tackle larger social factors. 2011.

[Pati2016] Elizabeth Patitsas, Jesse Berlin, Michelle Craig, and Steve Easterbrook. "Evidence that Computer Science Grades are not Bimodal." In: *2016 International Computing Education Research Conference (ICER'16)*. Presents a statistical analysis and an experiment which jointly show that grades in computing classes are not bimodal. Association for Computing Machinery (ACM), 2016. DOI: 10.1145/2960310.2960312.

[**Pea1986**] Roy D. Pea. "Language-Independent Conceptual "Bugs" in Novice Programming." In: *Journal of Educational Computing Research* 2.1 (Feb. 1986). First named the "superbug" in coding: most newcomers think the computer understands what they want, in the same way that a human being would, pp. 25–36. DOI: 10.2190/689t-1r2a-x4w4-29j2.

[**Petr2016**] Marian Petre and André van der Hoek. "Software Design Decoded: 66 Ways Experts Think." A short illustrated overview of how expert software developers think. MIT Press, 2016. ISBN: 0262035189.

[**Pign2016**] Alessandra Pigni. "The Idealist's Survival Kit: 75 Simple Ways to Prevent Burnout." A guide to staying sane and healthy while doing good. Parallax Press, 2016. ISBN: 1941529348.

[**Port2013**] Leo Porter, Mark Guzdial, Charlie McDowell, and Beth Simon. "Success in Introductory Programming: What Works?." In: *Communications of the ACM* 56.8 (Aug. 2013). Summarizes the evidence that peer instruction, media computation, and pair programming can significantly improve outcomes in introductory programming courses, p. 34. DOI: 10.1145/2492007.2492020.

[**Port2016**] Leo Porter, Dennis Bouvier, Quintin Cutts, Scott Grissom, Cynthia Bailey Lee, Robert McCartney, Daniel Zingaro, and Beth Simon. "A Multi-Institutional Study of Peer Instruction in Introductory Computing." In: *2016 Technical Symposium on Computer Science Education (SIGCSE'16)*. Reports that students in introductory programming classes value peer instruction, and that it improves learning outcomes. Association for Computing Machinery (ACM), 2016. DOI: 10.1145/2839509.2844642.

[**Qian2017**] Yizhou Qian and James Lehman. "Students' Misconceptions and Other Difficulties in Introductory Programming." In: *ACM Transactions on Computing Education* 18.1 (Oct. 2017). Summarizes research on student misconceptions about computing, pp. 1–24. DOI: 10.1145/3077618.

[**Rago2017**] Noa Ragonis and Ronit Shmallo. "On the (Mis)understanding of the this Reference." In: *2017 Technical Symposium on Computer Science Education (SIGCSE'17)*. Reports that most students do not understood when to use this, and that teachers are also often not clear on the subject. Association for Computing Machinery (ACM), 2017. DOI: 10.1145/3017680.3017715.

[**Rams2019**] G. Ramsay, A. B. Haynes, S. R. Lipsitz, I. Solsky, J. Leitch, A. A. Gawande, and M. Kumar. "Reducing surgical mortality in Scotland by use of the WHO Surgical Safety Checklist." In: *BJS* (Apr. 2019). Found that the introduction of surgical checklists in Scottish hospitals significantly reduced mortality rates. DOI: 10.1002/bjs.11151.

[**Raws2014**] Katherine A. Rawson, Ruthann C. Thomas, and Larry L. Jacoby. "The Power of Examples: Illustrative Examples Enhance Conceptual Learning of Declarative Concepts." In: *Educational Psychology Review* 27.3 (June 2014).

Reports that presenting examples helps students understand definitions, so long as examples and definitions are interleaved, pp. 483–504. DOI: 10.1007/s10648-014-9273-3.

[Ray2014] Eric J. Ray and Deborah S. Ray. "Unix and Linux: Visual QuickStart Guide." Fifth. An introduction to Unix that is both a good tutorial and a good reference guide. Peachpit Press, 2014. ISBN: 0321997549.

[Rice2018] Gail Taylor Rice. "Hitting Pause: 65 Lecture Breaks to Refresh and Reinforce Learning." Justifies and catalogs ways to take a pause in class to help learning. Stylus Publishing, 2018. ISBN: 9781620366530.

[Rich2017] Kathryn M. Rich, Carla Strickland, T. Andrew Binkowski, Cheryl Moran, and Diana Franklin. "K-8 learning Trajectories Derived from Research Literature." In: *2017 International Computing Education Research Conference (ICER'17)*. Presents learning trajectories for K-8 computing classes for Sequence, Repetition, and Conditions gleaned from the literature. Association for Computing Machinery (ACM), 2017. DOI: 10.1145/3105726.3106166.

[Ritz2018] Anna Ritz. "Programming the Central Dogma: An Integrated Unit on Computer Science and Molecular Biology Concepts." In: *2018 Technical Symposium on Computer Science Education (SIGCSE'18)*. Describes an introductory computing course for biologists whose problems are drawn from the DNA-to-protein processes in cells. Association for Computing Machinery (ACM), 2018. DOI: 10.1145/3159450.3159590.

[Robe2017] Eric Roberts. *Assessing and Responding to the Growth of Computer Science Undergraduate Enrollments: Annotated Findings.* http://cs.stanford.edu/people/eroberts/ResourcesForTheCSCapacityCrisis/files/AnnotatedFindings.pptx. Summarizes findings from a National Academies study about computer science enrollments. 2017.

[Robi2005] Evan Robinson. *Why Crunch Mode Doesn't Work: 6 Lessons.* http://www.igda.org/articles/erobinson_crunch.php. Summarizes research on the effects of overwork and sleep deprivation. 2005.

[Roge2018] Steven G. Rogelberg. "The Surprising Science of Meetings." A short summary of research on effective meetings. Oxford University Press, 2018. ISBN: 978-0190689216.

[Rohr2015] Doug Rohrer, Robert F. Dedrick, and Sandra Stershic. "Interleaved Practice Improves Mathematics Learning." In: *Journal of Educational Psychology* 107.3 (2015). Reports that interleaved practice is more effective than monotonous practice when learning, pp. 900–908. DOI: 10.1037/edu0000001.

[Rubi2013] Marc J. Rubin. "The Effectiveness of Live-coding to Teach Introductory Programming." In: *2013 Technical Symposium on Computer Science Education (SIGCSE'13)*. Reports that live coding is as good as or better than using static code examples. Association for Computing Machinery (ACM), 2013, pp. 651–656. DOI: 10.1145/2445196.2445388.

[**Rubi2014**] Manuel Rubio-Sánchez, Päivi Kinnunen, Cristóbal Pareja-Flores, and J. Ángel Velázquez-Iturbide. "Student Perception and Usage of an Automated Programming Assessment Tool." In: *Computers in Human Behavior* 31 (Feb. 2014). Describes use of an auto-grader for student assignments, pp. 453–460. DOI: 10.1016/j.chb.2013.04.001.

[**Sahl2015**] Pasi Sahlberg. "Finnish Lessons 2.0: What Can the World Learn from Educational Change in Finland?." A frank look at the success of Finland's educational system and why other countries struggle to replicate it. Teachers College Press, 2015. ISBN: 978-0807755853.

[**Saja2006**] Jorma Sajaniemi, Mordechai Ben-Ari, Pauli Byckling, Petri Gerdt, and Yevgeniya Kulikova. "Roles of Variables in Three Programming Paradigms." In: *Computer Science Education* 16.4 (Dec. 2006). A detailed look at the authors' work on roles of variables, pp. 261–279. DOI: 10 . 1080 / 08993400600874584.

[**Sala2017**] Giovanni Sala and Fernand Gobet. "Does Far Transfer Exist? Negative Evidence From Chess, Music, and Working Memory Training." In: *Current Directions in Psychological Science* 26.6 (Oct. 2017). A meta-analysis showing that far transfer rarely occurs, pp. 515–520. DOI: 10.1177/0963721417712760.

[**Sand2013**] Kate Sanders, Jaime Spacco, Marzieh Ahmadzadeh, Tony Clear, Stephen H. Edwards, Mikey Goldweber, Chris Johnson, Raymond Lister, Robert McCartney, and Elizabeth Patitsas. "The Canterbury QuestionBank: Building a Repository of Multiple-Choice CS1 and CS2 Questions." In: *2013 Conference on Innovation and Technology in Computer Science Education (ITiCSE'13)*. Describes development of a shared question bank for introductory CS, and patterns for multiple choice questions that emerged from entries. Association for Computing Machinery (ACM), 2013. DOI: 10.1145/2543882.2543885.

[**Scan1989**] David A. Scanlan. "Structured Flowcharts Outperform Pseudocode: An Experimental Comparison." In: *IEEE Software* 6.5 (Sept. 1989). Reports that students understand flowcharts better than pseudocode if both are equally well structured, pp. 28–36. DOI: 10.1109/52.35587.

[**Scho1984**] Donald A. Schön. "The Reflective Practitioner: How Professionals Think In Action." A groundbreaking look at how professionals in different fields actually solve problems. Basic Books, 1984. ISBN: 0465068782.

[**Schw2013**] Viviane Schwarz. "Welcome to Your Awesome Robot." A wonderful illustrated guide to building wearable cardboard robot suits. Not just for kids. Flying Eye Books, 2013. ISBN: 978-1909263000.

[**Scot1987**] James C. Scott. "Weapons of the Weak: Everyday Forms of Peasant Resistance." Describes the techniques of evasion and resistance that the weak use to resist the strong. Yale University Press, 1987. ISBN: 978-0300036411.

[Scot1998] James C. Scott. "Seeing Like a State: How Certain Schemes to Improve the Human Condition Have Failed." Argues that large organizations consistently prefer uniformity over productivity. Yale University Press, 1998. ISBN: 0300078153.

[Sent2018] Sue Sentance, Erik Barendsen, and Carsten Schulte, eds. "Computer Science Education: Perspectives on Teaching and Learning in School." A collection of academic survey articles on teaching computing. Bloomsbury Press, 2018. ISBN: 135005710X.

[Sent2019] Sue Sentance, Jane Waite, and Maria Kallia. "Teachers' Experiences of using PRIMM to Teach Programming in School." In: *2019 Technical Symposium on Computer Science Education (SIGCSE'19)*. Describes PRIMM and its effectiveness. ACM Press, 2019. DOI: 10.1145/3287324.3287477.

[Sepp2015] Otto Seppälä, Petri Ihantola, Essi Isohanni, Juha Sorva, and Arto Vihavainen. "Do We Know How Difficult the Rainfall Problem Is?." In: *2015 Koli Calling Conference on Computing Education Research (Koli'15)*. A meta-study of the Rainfall Problem. ACM Press, 2015. DOI: 10.1145/2828959.2828963.

[Shap2007] Jenessa R. Shapiro and Steven L. Neuberg. "From Stereotype Threat to Stereotype Threats: Implications of a Multi-Threat Framework for Causes, Moderators, Mediators, Consequences, and Interventions." In: *Personality and Social Psychology Review* 11.2 (May 2007). Explores the ways the term "stereotype threat" has been used, pp. 107–130. DOI: 10.1177/1088868306294790.

[Shel2017] Duane F. Shell, Leen-Kiat Soh, Abraham E. Flanigan, Markeya S. Peteranetz, and Elizabeth Ingraham. "Improving Students' Learning and Achievement in CS Classrooms Through Computational Creativity Exercises that Integrate Computational and Creative Thinking." In: *2017 Technical Symposium on Computer Science Education (SIGCSE'17)*. Reports that having students work in small groups on computational creativity exercises improves learning outcomes. Association for Computing Machinery (ACM), 2017. DOI: 10.1145/3017680.3017718.

[Shol2019] Dan Sholler, Igor Steinmacher, Denae Ford, Mara Averick, Mike Hoye, and Greg Wilson. *Ten Simple Rules for Helping Newcomers Become Contributors to Open Source Projects*. https://github.com/gvwilson/10-newcomers/. Evidence-based practices for helping newcomers become productive in open projects. 2019.

[Simo2013] Simon. "Soloway's Rainfall Problem has Become Harder." In: *2013 Conference on Learning and Teaching in Computing and Engineering*. Argues that the Rainfall problem is harder for novices than it used to be because they're not used to handling keyboard input, so direct comparison with past results may be unfair. Institute of Electrical and Electronics Engineers (IEEE), Mar. 2013. DOI: 10.1109/latice.2013.44.

[Sing2012] Vandana Singh. "Newcomer integration and learning in technical support communities for open source software." In: *2012 ACM International Conference on Supporting Group Work - GROUP'12*. An early study of onboarding in open source. ACM Press, 2012. DOI: 10.1145/2389176.2389186.

[Sirk2012] Teemu Sirkiä and Juha Sorva. "Exploring Programming Misconceptions: An Analysis of Student Mistakes in Visual Program Simulation Exercises." In: *2012 Koli Calling Conference on Computing Education Research (Koli'12)*. Analyzes data from student use of an execution visualization tool and classifies common mistakes. Association for Computing Machinery (ACM), 2012. DOI: 10.1145/2401796.2401799.

[Sisk2018] Victoria F. Sisk, Alexander P. Burgoyne, Jingze Sun, Jennifer L. Butler, and Brooke N. Macnamara. "To What Extent and Under Which Circumstances Are Growth Mind-Sets Important to Academic Achievement? Two Meta-Analyses." In: *Psychological Science* (Mar. 2018). Reports meta-analyses of the relationship between mind-set and academic achievement, and the effectiveness of mind-set interventions on academic achievement, and finds that overall effects are weak for both, but some results support specific tenets of the theory, p. 095679761773970. DOI: 10.1177/0956797617739704.

[Skud2014] Ben Skudder and Andrew Luxton-Reilly. "Worked Examples in Computer Science." In: *2014 Australasian Computing Education Conference, (ACE'14)*. A summary of research on worked examples as applied to computing education. 2014.

[Smar2018] Benjamin L. Smarr and Aaron E. Schirmer. "3.4 Million Real-World Learning Management System Logins Reveal the Majority of Students Experience Social Jet Lag Correlated with Decreased Performance." In: *Scientific Reports* 8.1 (Mar. 2018). Reports that students who have to work outside their natural body clock cycle do less well. DOI: 10.1038/s41598-018-23044-8.

[Smit2009] Michelle K. Smith, William B. Wood, Wendy K. Adams, Carl E. Wieman, Jennifer K. Knight, N. Guild, and T. T. Su. "Why Peer Discussion Improves Student Performance on In-class Concept Questions." In: *Science* 323.5910 (Jan. 2009). Reports that student understanding increases during discussion in peer instruction, even when none of the students in the group initially know the right answer, pp. 122–124. DOI: 10.1126/science.1165919.

[Solo1984] Elliot Soloway and Kate Ehrlich. "Empirical Studies of Programming Knowledge." In: *IEEE Transactions on Software Engineering* SE-10.5 (Sept. 1984). Proposes that experts have programming plans and rules of programming discourse, pp. 595–609. DOI: 10.1109/tse.1984.5010283.

[Solo1986] Elliot Soloway. "Learning to Program = Learning to Construct Mechanisms and Explanations." In: *Communications of the ACM* 29.9 (Sept. 1986). Analyzes programming in terms of choosing appropriate goals and constructing plans to achieve them, and introduces the Rainfall Problem, pp. 850–858. DOI: 10.1145/6592.6594.

[Sond2012] Harald Søndergaard and Raoul A. Mulder. "Collaborative Learning Through Formative Peer Review: Pedagogy, Programs and Potential." In: *Computer Science Education* 22.4 (Dec. 2012). Surveys literature on student peer assessment, distinguishing grading and reviewing as separate forms, and summarizes features a good peer review system needs to have, pp. 343–367. DOI: 10.1080/08993408.2012.728041.

[Sorv2013] Juha Sorva. "Notional Machines and Introductory Programming Education." In: *ACM Transactions on Computing Education* 13.2 (June 2013). Reviews literature on programming misconceptions, and argues that instructors should address notional machines as an explicit learning objective, pp. 1–31. DOI: 10.1145/2483710.2483713.

[Sorv2014] Juha Sorva and Otto Seppälä. "Research-based Design of the First Weeks of CS1." In: *2014 Koli Calling Conference on Computing Education Research (Koli'14)*. Proposes three cognitively plausible frameworks for the design of a first CS course. Association for Computing Machinery (ACM), 2014. DOI: 10.1145/2674683.2674690.

[Sorv2018] Juha Sorva. "Misconceptions and the Beginner Programmer." In: *Computer Science Education: Perspectives on Teaching and Learning in School*. Ed. by Sue Sentance, Erik Barendsen, and Carsten Schulte. Summarizes what we know about what novices misunderstand about computing. Bloomsbury Press, 2018. ISBN: 135005710X.

[Spal2014] Dan Spalding. "How to Teach Adults: Plan Your Class, Teach Your Students, Change the World." A short guide to teaching adult free-range learners informed by the author's social activism. Jossey-Bass, 2014. ISBN: 1118841360.

[Spoh1985] James C. Spohrer, Elliot Soloway, and Edgar Pope. "A Goal/Plan Analysis of Buggy Pascal Programs." In: *Human-Computer Interaction* 1.2 (June 1985). One of the first cognitively plausible analyses of how people program, which proposes a goal/plan model, pp. 163–207. DOI: 10.1207/s15327051hci0102_4.

[Srid2016] Sumukh Sridhara, Brian Hou, Jeffrey Lu, and John DeNero. "Fuzz Testing Projects in Massive Courses." In: *2016 Conference on Learning @ Scale (L@S'16)*. Reports that fuzz testing student code catches errors that are missed by handwritten test suite, and explains how to safely share tests and results. Association for Computing Machinery (ACM), 2016. DOI: 10.1145/2876034.2876050.

[Stam2013] Eliane Stampfer and Kenneth R. Koedinger. "When Seeing Isn't Believing: Influences of Prior Conceptions and Misconceptions." In: *2013 Annual Meeting of the Cognitive Science Society (CogSci'13)*. Explores why giving children more information when they are learning about fractions can lower their performance. 2013.

[**Stam2014**] Eliane Stampfer Wiese and Kenneth R. Koedinger. "Investigating Scaffolds for Sense Making in Fraction Addition and Comparison." In: *2014 Annual Conference of the Cognitive Science Society (CogSci'14)*. Looks at how to scaffold learning of fraction operations. 2014.

[**Star2014**] Philip Stark and Richard Freishtat. "An Evaluation of Course Evaluations." In: *ScienceOpen Research* (Sept. 2014). Yet another demonstration that teaching evaluations don't correlate with learning outcomes, and that they are frequently statistically suspect. DOI: 10.14293/s2199-1006.1.sor-edu.aofrqa. v1.

[**Stas1998**] John Stasko, John Domingue, Mark H. Brown, and Blaine A. Price, eds. "Software Visualization: Programming as a Multimedia Experience." A survey of program and algorithm visualization techniques and results. MIT Press, 1998. ISBN: 0262193957.

[**Stee2011**] Claude M. Steele. "Whistling Vivaldi: How Stereotypes Affect Us and What We Can Do." Explains and explores stereotype threat and strategies for addressing it. W. W. Norton & Company, 2011. ISBN: 0393339726.

[**Stef2013**] Andreas Stefik and Susanna Siebert. "An Empirical Investigation into Programming Language Syntax." In: *ACM Transactions on Computing Education* 13.4 (Nov. 2013). Reports that curly-brace languages are as hard to learn as a language with randomly designed syntax, but others are easier, pp. 1–40. DOI: 10.1145/2534973.

[**Stef2017**] Andreas Stefik, Patrick Daleiden, Diana Franklin, Stefan Hanenberg, Antti-Juhani Kaijanaho, Walter Tichy, and Brett A. Becker. *Programming Languages and Learning*. https://quorumlanguage.com/evidence.html. Summarizes what we actually know about designing programming languages and why we believe it's true. 2017.

[**Steg2014**] Martijn Stegeman, Erik Barendsen, and Sjaak Smetsers. "Towards an Empirically Validated Model for Assessment of Code Quality." In: *2014 Koli Calling Conference on Computing Education Research (Koli'14)*. Presents a code quality rubric for novice programming courses. Association for Computing Machinery (ACM), 2014. DOI: 10.1145/2674683.2674702.

[**Steg2016a**] Martijn Stegeman, Erik Barendsen, and Sjaak Smetsers. "Designing a Rubric for Feedback on Code Quality in Programming Courses." In: *2016 Koli Calling Conference on Computing Education Research (Koli'16)*. Describes several iterations of a code quality rubric for novice programming courses. Association for Computing Machinery (ACM), 2016. DOI: 10.1145/2999541. 2999555.

[**Steg2016b**] Martijn Stegeman, Erik Barendsen, and Sjaak Smetsers. *Rubric for Feedback on Code Quality in Programming Courses*. http://stgm.nl/quality. Presents a code quality rubric for novice programming. 2016.

[**Stei2013**] Igor Steinmacher, Igor Wiese, Ana Paula Chaves, and Marco Aurelio Gérosa. "Why do newcomers abandon open source software projects?." In:

2013 International Workshop on Cooperative and Human Aspects of Software Engineering (CHASE'13). Explores why new members *don't* stay in open source projects. Institute of Electrical and Electronics Engineers (IEEE), May 2013. DOI: 10.1109/chase.2013.6614728.

[**Stei2016**] Igor Steinmacher, Tayana Uchoa Conte, Christoph Treude, and Marco Aurélio Gerosa. "Overcoming open source project entry barriers with a portal for newcomers." In: *2016 International Conference on Software Engineering (ICSE'16)*. Reports the effectiveness of a portal specifically designed to help newcomers. ACM Press, 2016. DOI: 10.1145/2884781.2884806.

[**Stei2018**] Igor Steinmacher, Gustavo Pinto, Igor Scaliante Wiese, and Marco Aurélio Gerosa. "Almost There: A Study on Quasi-Contributors in Open-Source Software Projects." In: *2018 International Conference on Software Engineering (ICSE'18)*. Look at why external developers fail to get their contributions accepted into open source projects. ACM Press, 2018. DOI: 10.1145/3180155.3180208.

[**Stoc2018**] Jean Stockard, Timothy W. Wood, Cristy Coughlin, and Caitlin Rasplica Khoury. "The Effectiveness of Direct Instruction Curricula: A Meta-analysis of a Half Century of Research." In: *Review of Educational Research* (Jan. 2018). A meta-analysis that finds significant positive benefit for Direct Instruction, p. 003465431775191. DOI: 10.3102/0034654317751919.

[**Sung2012**] Eunmo Sung and Richard E. Mayer. "When Graphics Improve Liking but not Learning from Online Lessons." In: *Computers in Human Behavior* 28.5 (Sept. 2012). Reports that students who receive any kind of graphics give significantly higher satisfaction ratings than those who don't, but only students who get instructive graphics perform better than groups that get no graphics, seductive graphics, or decorative graphics, pp. 1618–1625. DOI: 10.1016/j.chb.2012.03.026.

[**Sved2016**] Maria Svedin and Olle Bälter. "Gender Neutrality Improved Completion Rate for All." In: *Computer Science Education* 26.2-3 (July 2016). Reports that redesigning an online course to be gender neutral improves completion probability in general, but decreases it for students with a superficial approach to learning, pp. 192–207. DOI: 10.1080/08993408.2016.1231469.

[**Tedr2008**] Matti Tedre and Erkki Sutinen. "Three Traditions of Computing: What Educators Should Know." In: *Computer Science Education* 18.3 (Sept. 2008). Summarizes the history and views of three traditions in computing: mathematical, scientific, and engineering, pp. 153–170. DOI: 10.1080/08993400802332332.

[**Tew2011**] Allison Elliott Tew and Mark Guzdial. "The FCS1: A Language Independent Assessment of CS1 Knowledge." In: *2011 Technical Symposium on Computer Science Education (SIGCSE'11)*. Describes development and validation of a language-independent assessment instrument for CS1 knowledge. Association for Computing Machinery (ACM), 2011. DOI: 10.1145/1953163.1953200.

[**Thay2017**] Kyle Thayer and Andrew J. Ko. "Barriers Faced by Coding Bootcamp Students." In: *2017 International Computing Education Research Conference (ICER'17)*. Reports that coding bootcamps are sometimes useful, but quality is varied, and formal and informal barriers to employment remain. Association for Computing Machinery (ACM), 2017. DOI: 10.1145/3105726.3106176.

[**Ubel2017**] Robert Ubell. *How the Pioneers of the MOOC got It Wrong*. http://spectrum.ieee.org/tech-talk/at-work/education/how-the-pioneers-of-the-mooc-got-it-wrong. A brief exploration of why MOOCs haven't lived up to initial hype. 2017.

[**Urba2014**] David R. Urbach, Anand Govindarajan, Refik Saskin, Andrew S. Wilton, and Nancy N. Baxter. "Introduction of Surgical Safety Checklists in Ontario, Canada." In: *New England Journal of Medicine* 370.11 (Mar. 2014). Reports a study showing that the introduction of surgical checklists did not have a significant effect on operative outcomes, pp. 1029–1038. DOI: 10.1056/nejmsa1308261.

[**Utti2013**] Ian Utting, Juha Sorva, Tadeusz Wilusz, Allison Elliott Tew, Michael McCracken, Lynda Thomas, Dennis Bouvier, Roger Frye, James Paterson, Michael E. Caspersen, and Yifat Ben-David Kolikant. "A Fresh Look at Novice Programmers' Performance and Their Teachers' Expectations." In: *2013 Conference on Innovation and Technology in Computer Science Education (ITiCSE'13)*. Replicates an earlier study showing how little students learn in their first programming course. ACM Press, 2013. DOI: 10.1145/2543882.2543884.

[**Uttl2017**] Bob Uttl, Carmela A. White, and Daniela Wong Gonzalez. "Meta-analysis of Faculty's Teaching Effectiveness: Student Evaluation of Teaching Ratings and Student Learning are not Related." In: *Studies in Educational Evaluation* 54 (Sept. 2017). Summarizes studies showing that how students rate a course and how much they actually learn are not related, pp. 22–42. DOI: 10.1016/j.stueduc.2016.08.007.

[**Varm2015**] Roli Varma and Deepak Kapur. "Decoding Femininity in Computer Science in India." In: *Communications of the ACM* 58.5 (Apr. 2015). Reports female participation in computing in India, pp. 56–62. DOI: 10.1145/2663339.

[**Vell2017**] Mickey Vellukunnel, Philip Buffum, Kristy Elizabeth Boyer, Jeffrey Forbes, Sarah Heckman, and Ketan Mayer-Patel. "Deconstructing the Discussion Forum: Student Questions and Computer Science Learning." In: *2017 Technical Symposium on Computer Science Education (SIGCSE'17)*. Found that students mostly ask constructivist and logistical questions in forums, and that the former correlate with grades. Association for Computing Machinery (ACM), 2017. DOI: 10.1145/3017680.3017745.

[**Viha2014**] Arto Vihavainen, Jonne Airaksinen, and Christopher Watson. "A Systematic Review of Approaches for Teaching Introductory Programming and Their Influence on Success." In: *2014 International Computing Education Research Conference (ICER'14)*. Consolidates studies of CS1-level teaching changes and finds media computation the most effective, while introducing a game theme is the least effective. Association for Computing Machinery (ACM), 2014. DOI: 10.1145/2632320.2632349.

[Wall2009] Thorbjorn Walle and Jo Erskine Hannay. "Personality and the Nature of Collaboration in Pair Programming." In: *2009 International Symposium on Empirical Software Engineering and Measurement (ESER'09)*. Reports that pairs with different levels of a given personality trait communicated more intensively. Institute of Electrical and Electronics Engineers (IEEE), Oct. 2009. DOI: 10.1109/esem.2009.5315996.

[Wang2018] April Y. Wang, Ryan Mitts, Philip J. Guo, and Parmit K. Chilana. "Mismatch of Expectations: How Modern Learning Resources Fail Conversational Programmers." In: *2018 Conference on Human Factors in Computing Systems (CHI'18)*. Reports that learning resources don't really help conversational programmers (those who learn coding to take part in technical discussions). Association for Computing Machinery (ACM), 2018. DOI: 10.1145/3173574.3174085.

[Ward2015] James Ward. "Adventures in Stationery: A Journey Through Your Pencil Case." A wonderful look at the everyday items that would be in your desk drawer if someone hadn't walked off with them. Profile Books, 2015. ISBN: 1846686164.

[Wats2014] Christopher Watson and Frederick W. B. Li. "Failure Rates in Introductory Programming Revisited." In: *2014 Conference on Innovation and Technology in Computer Science Education (ITiCSE'14)*. A larger version of an earlier study that found an average of one third of students fail CS1. Association for Computing Machinery (ACM), 2014. DOI: 10.1145/2591708.2591749.

[Watt2014] Audrey Watters. "The Monsters of Education Technology." A collection of essays about the history of educational technology and the exaggerated claims repeatedly made for it. CreateSpace, 2014. ISBN: 1505225051.

[Wein2017] David Weintrop and Uri Wilensky. "Comparing Block-Based and Text-Based Programming in High School Computer Science Classrooms." In: *ACM Transactions on Computing Education* 18.1 (Oct. 2017). Reports that students learn faster and better with blocks than with text, pp. 1–25. DOI: 10.1145/3089799.

[Wein2018a] Yana Weinstein, Christopher R. Madan, and Megan A. Sumeracki. "Teaching the Science of Learning." In: *Cognitive Research: Principles and Implications* 3.1 (Jan. 2018). A tutorial review of six evidence-based learning practices. DOI: 10.1186/s41235-017-0087-y.

[Wein2018b] Yana Weinstein, Megan Sumeracki, and Oliver Caviglioli. "Understanding How We Learn: A Visual Guide." A short graphical summary of effective learning strategies. Routledge, 2018. ISBN: 978-1138561724.

[Weng2015] Etienne Wenger-Trayner and Beverly Wenger-Trayner. *Communities of Practice: A Brief Introduction.* http://wenger-trayner.com/intro-to-cops/. A brief summary of what communities of practice are and aren't. 2015.

[Wibu2016] Karin Wiburg, Julia Parra, Gaspard Mucundanyi, Jennifer Green, and Nate Shaver, eds. "The Little Book of Learning Theories." Second. Presents brief summaries of various theories of learning. CreateSpace, 2016. ISBN: 1537091808.

[Wigg2005] Grant Wiggins and Jay McTighe. "Understanding by Design." A lengthy presentation of reverse instructional design. Association for Supervision & Curriculum Development (ASCD), 2005. ISBN: 1416600353.

[Wilc2018] Chris Wilcox and Albert Lionelle. "Quantifying the Benefits of Prior Programming Experience in an Introductory Computer Science Course." In: *2018 Technical Symposium on Computer Science Education (SIGCSE'18).* Reports that students with prior experience outscore students without in CS1, but there is no significant difference in performance by the end of CS2; also finds that female students with prior exposure outperform their male peers in all areas, but are consistently less confident in their abilities. Association for Computing Machinery (ACM), 2018. DOI: 10.1145/3159450.3159480.

[Wile2002] David Wiley. *The Reusability Paradox.* http://opencontent.org/docs/paradox.html. Summarizes the tension between learnings objects being effective and reusable. 2002.

[Wilk2011] Richard Wilkinson and Kate Pickett. "The Spirit Level: Why Greater Equality Makes Societies Stronger." Presents evidence that inequality harms everyone, both economically and otherwise. Bloomsbury Press, 2011. ISBN: 1608193411.

[Will2010] Daniel T. Willingham. "Why Don't Students Like School?: A Cognitive Scientist Answers Questions about How the Mind Works and What It Means for the Classroom." A cognitive scientist looks at how the mind works in the classroom. Jossey-Bass, 2010. ISBN: 047059196X.

[Wils2007] Karen Wilson and James H. Korn. "Attention During Lectures: Beyond Ten Minutes." In: *Teaching of Psychology* 34.2 (June 2007). Reports little support for the claim that students only have a 10–15 minute attention span (though there is lots of individual variation), pp. 85–89. DOI: 10.1080/00986280701291291.

[Wils2016] Greg Wilson. "Software Carpentry: Lessons Learned." In: *F1000Research* (Jan. 2016). A history and analysis of Software Carpentry. DOI: 10.12688/f1000research.3-62.v2.

[Wlod2017] Raymond J. Wlodkowski and Margery B. Ginsberg. "Enhancing Adult Motivation to Learn: A Comprehensive Guide for Teaching All Adults." The standard reference for understanding adult motivation. Jossey-Bass, 2017. ISBN: 1119077990.

[Xie2019] Benjamin Xie, Dastyni Loksa, Greg L. Nelson, Matthew J. Davidson, Dongsheng Dong, Harrison Kwik, Alex Hui Tan, Leanne Hwa, Min Li, and Andrew J. Ko. "A theory of instruction for introductory programming skills." In: *Computer Science Education* 29.2-3 (Jan. 2019). Lays out a four-part theory for teaching novices based on reading vs. writing and code vs. templates, pp. 205–253. DOI: 10.1080/08993408.2019.1565235.

[Yada2016] Aman Yadav, Sarah Gretter, Susanne Hambrusch, and Phil Sands. "Expanding Computer Science Education in Schools: Understanding Teacher Experiences and Challenges." In: *Computer Science Education* 26.4 (Dec. 2016). Summarizes feedback from K-12 teachers on what they need by way of preparation and support, pp. 235–254. DOI: 10.1080/08993408.2016.1257418.

[Yang2015] Yu-Fen Yang and Yuan-Yu Lin. "Online Collaborative Note-Taking Strategies to Foster EFL Beginners' Literacy Development." In: *System* 52 (Aug. 2015). Reports that students using collaborative note taking when learning English as a foreign language do better than those who don't, pp. 127–138. DOI: 10.1016/j.system.2015.05.006.

A License

This is a human-readable summary of (and not a substitute for) the license. Please see https://creativecommons.org/licenses/by-nc/4.0/legalcode for the full legal text.

This work is licensed under the Creative Commons Attribution-NonCommercial 4.0[1] license (CC-BY-NC-4.0).

You are free to:

- **Share**—copy and redistribute the material in any medium or format
- **Adapt**—remix, transform, and build upon the material.

The licensor cannot revoke these freedoms as long as you follow the license terms.

Under the following terms:

- **Attribution**—You must give appropriate credit, provide a link to the license, and indicate if changes were made. You may do so in any reasonable manner, but not in any way that suggests the licensor endorses you or your use.

- **NonCommercial**—You may not use the material for commercial purposes.

No additional restrictions—You may not apply legal terms or technological measures that legally restrict others from doing anything the license permits.

Notices:

- You do not have to comply with the license for elements of the material in the public domain or where your use is permitted by an applicable exception or limitation.
- No warranties are given. The license may not give you all of the permissions necessary for your intended use. For example, other rights such as publicity, privacy, or moral rights may limit how you use the material.

---

[1] https://creativecommons.org/licenses/by-nc/4.0/

B Code of Conduct

In the interest of fostering an open and welcoming environment, we as contributors and maintainers pledge to making participation in our project and our community a harassment-free experience for everyone, regardless of age, body size, disability, ethnicity, gender identity and expression, level of experience, education, socioeconomic status, nationality, personal appearance, race, religion, or sexual identity and orientation.

OUR STANDARDS

Examples of behavior that contribute to creating a positive environment include:

- using welcoming and inclusive language,
- being respectful of differing viewpoints and experiences,
- gracefully accepting constructive criticism,
- focusing on what is best for the community, and
- showing empathy towards other community members.

Examples of unacceptable behavior by participants include:

- the use of sexualized language or imagery and unwelcome sexual attention or advances,
- trolling, insulting/derogatory comments, and personal or political attacks,
- public or private harassment,
- publishing others' private information, such as a physical or electronic address, without explicit permission, and
- other conduct which could reasonably be considered inappropriate in a professional setting.

OUR RESPONSIBILITIES

Project maintainers are responsible for clarifying the standards of acceptable behavior and are expected to take appropriate and fair corrective action in response to any instances of unacceptable behavior.

Project maintainers have the right and responsibility to remove, edit, or reject comments, commits, code, wiki edits, issues, and other contributions that are not aligned to this Code of Conduct, or to ban temporarily or permanently any contributor for other behaviors that they deem inappropriate, threatening, offensive, or harmful.

SCOPE

This Code of Conduct applies both within project spaces and in public spaces when an individual is representing the project or its community. Examples of representing a project or community include using an official project e-mail address, posting via an official social media account, or acting as an appointed representative at an online or offline event. Representation of a project may be further defined and clarified by project maintainers.

ENFORCEMENT

Instances of abusive, harassing, or otherwise unacceptable behavior may be reported by emailing the project's maintainer at gvwilson@third-bit.com. All complaints will be reviewed and investigated and will result in a response that is deemed necessary and appropriate to the circumstances. The project team is obligated to maintain confidentiality with regard to the reporter of an incident. Further details of specific enforcement policies may be posted separately.

Project maintainers who do not follow or enforce the Code of Conduct in good faith may face temporary or permanent repercussions as determined by other members of the project's leadership.

ATTRIBUTION

This Code of Conduct is adapted from the Contributor Covenant[1] version 1.4.

---

[1] https://www.contributor-covenant.org

C Joining Our Community

We hope you will choose to help us do the same for this book. If you are new to working this way, please see Appendix B for our code of conduct, and then:

Start small. Fix a typo, clarify the wording of an exercise, correct or update a citation, or suggest a better example or analogy to illustrate some point.

Join the conversation. Have a look at the issues and proposed changes that other people have already filed and add your comments to them. It's often possible to improve improvements, and it's a good way to introduce yourself to the community and make new friends.

Discuss, then edit. If you want to propose a large change, such as reorganizing or splitting an entire chapter, please file an issue that outlines your proposal and your reasoning and tag it with "Proposal." We encourage everyone to add comments to these issues so that the whole discussion of what and why is in the open and can be archived. If the proposal is accepted, the actual work may then be broken down into several smaller issues or changes that can be tackled independently.

C.1 USING THIS MATERIAL

As Chapter 1 stated, all of this material can be freely distributed and re-used. under the Creative Commons Attribution-NonCommercial 4.0 license (Appendix A). You can use the online version at http://teachtogether.tech/ in any class (free or paid), and can quote short excerpts under fair use[1] provisions, but cannot republish large parts in commercial works without prior permission.

This material has been used in many ways, from a multi-week online class to an intensive in-person workshop. It's usually possible to cover large parts of Chapter 2 to Chapter 6, Chapter 8, and Chapter 10 in two long days.

IN PERSON

This is the most effective way to deliver this training, but also the most demanding. Participants are physically together. When they need to practice teaching in small groups, some or all of them go to nearby breakout spaces. Participants use their own tablets or laptops to view online material during the class and for shared note-taking (Section 9.7), and use pen and paper or whiteboards for other exercises. Questions and discussion are done aloud.

If you are teaching in this format, you should use sticky notes as status flags so that you can see who needs help, who has questions, and who's ready to move on

---

[1] https://en.wikipedia.org/wiki/Fair_use

(Section 9.8). You should also use them to distribute attention so that everyone gets a fair share of the teacher's time, and as minute cards to encourage learners to reflect on what they've just learned and to give you actionable feedback while you still have time to act on it.

ONLINE IN GROUPS

In this format, 10–40 learners are together in 2–6 groups of 4–12, but those groups are geographically distributed. Each group uses one camera and microphone to connect to the video call, rather than each person being on the call separately. Good audio matters more than good video: the better the audio, the more learners can communicate with the teacher and other rooms by voice rather than text.

The entire class does shared note-taking together, and also uses the shared notes for asking and answering questions. Having several dozen people try to talk on a call works poorly, so in most sessions, the teacher does the talking and learners respond through the note-taking tool's chat.

ONLINE AS INDIVIDUALS

The natural extension of being online in groups is to be online as individuals. As with online groups, the teacher will do most of the talking and learners will mostly participate via text chat. Good audio is once again more important than good video, and participants should use text chat to signal that they want to speak next (Appendix E).

Having participants online individually makes it more difficult to draw and share concept maps (Section 3.4) or give feedback on teaching (Section 8.5). Teachers should therefore rely more on exercises with written results that can be put in the shared notes, such as giving feedback on stock videos of people teaching.

MULTI-WEEK ONLINE

The class meets every week for an hour via video conferencing. Each meeting may be held twice to accommodate learners' time zones and schedules. Participants use shared note-taking as described above for online group classes, post homework online between classes, and comment on each other's work. In practice, comments are relatively rare: people strongly prefer to discuss material in the weekly meetings.

This was the first format used, and I no longer recommend it: while spreading the class out gives people time to reflect and tackle larger exercises, it also greatly increases the odds that they'll have to drop out because of other demands on their time.

C.2 CONTRIBUTING AND MAINTAINING

Contributions of all kinds are welcome, from suggestions for improvements to errata and new material. All contributors must abide by our Code of Conduct (Appendix B);

by submitting your work, you are agreeing that it may incorporated in either original or edited form and released under the same license as the rest of this material
(Appendix A). If your material is incorporated, we will add you to the acknowledg-
ments (Section 1.3) unless you request otherwise.

The source for this book is stored on GitHub at:

https://github.com/gvwilson/teachtogether.tech/

If you know how to use Git and GitHub and would like to change, fix, or add something, please submit a **pull request** that modifies the LaTeX source. If you would
like to preview your changes, please run make pdf or make html on the command
line.

If you want to report an error, ask a question, or make a suggestion, please file an
issue in the repository. You need to have a GitHub account in order to do this, but do
not need to know how to use Git.

If you do not wish to create a GitHub account, please email your contribution to
gvwilson@third-bit.com with either "T3" or "Teaching Tech Together" somewhere in the subject line. We will try to respond within a week.

Finally, we always enjoy hearing how people have used this material, and are
always grateful for more diagrams.

D Glossary

Absolute beginner Someone who has never encountered concepts or material before. The term is used in distinction to **false beginner**.

Active learning An approach to instruction in which learners engage with material through discussion, problem solving, case studies, and other activities that require them to reflect on and use new information in real time. See also **passive learning**.

Active teaching An approach to instruction in which the teacher acts on new information acquired from learners while teaching (e.g., by dynamically changing an example or rearranging the intended order of content). See also **passive teaching**.

Authentic task A task which contains important elements of things that learners would do in real (non-classroom situations). To be authentic, a task should require learners to construct their own answers rather than choose between provided answers, and to work with the same tools and data they would use in real life.

Automaticity The ability to do a task without concentrating on its low-level details.

Backward design An instructional design method that works backwards from a summative assessment to formative assessments and thence to lesson content.

Behaviorism A theory of learning whose central principle is stimulus and response, and whose goal is to explain behavior without recourse to internal mental states or other unobservables. See also **cognitivism**.

Bloom's Taxonomy A six-part hierarchical classification of understanding whose levels are *knowledge*, *comprehension*, *application*, *analysis*, *synthesis*, and *evaluation* that has been widely adopted. See also **Fink's Taxonomy**.

Brand The associations people have with a product's name or identity.

Calibrated peer review Having learners compare their reviews of sample work with a teacher's reviews before being allowed to review their peers' work.

Chunking The act of grouping related concepts together so that they can be stored and processed as a single unit.

Co-teaching Teaching with another teacher in the classroom.

Cognitive apprenticeship A theory of learning that emphasizes the process of a master passing on skills and insights situationally to an apprentice.

Cognitive load The mental effort needed to solve a problem. Cognitive load theory divides this into **intrinsic**, **germane**, and **extraneous** load, and holds that people learn fastest when germane and extraneous load are reduced.

Cognitivism A theory of learning that holds that mental states and processes can and must be included in models of learning. See also **behaviorism**.

Commons something managed jointly by a community according to rules they themselves have evolved and adopted.

Community of practice A self-perpetuating group of people who share and develop a craft such as knitters, musicians, or programmers. See also **legitimate peripheral participation**.

Community representation Using cultural capital to highlight learners' social identities, histories, and community networks in learning activities.

Competent practitioner Someone who can do normal tasks with normal effort under normal circumstances. See also **novice** and **expert**.

Computational integration Using computing to re-implement pre-existing cultural artifacts, e.g., creating variants of traditional designs using computer drawing tools.

Computational thinking Thinking about problem-solving in ways inspired by programming (though the term is used in many other ways).

Concept inventory A test designed to determine how well a learner understands a domain. Unlike most instructor-authored tests, concept inventories are based on extensive research and validation.

Concept map A picture of a mental model in which concepts are nodes in a graph and relationships are (labeled) arcs.

Connectivism A theory of learning holds that knowledge is distributed, that learning is the process of navigating, growing, and pruning connections, and which emphasizes the social aspects of learning made possible by the Internet.

Constructivism A theory of learning that views learners as actively constructing knowledge.

Content knowledge A person's understanding of a subject. See also **general pedagogical knowledge** and **pedagogical content knowledge**.

Contributing student pedagogy Having learners produce artifacts to contribute to others' learning.

Conversational programmer Someone who needs to know enough about computing to have a meaningful conversation with a programmer, but isn't going to program themselves.

CS Unplugged A style of teaching that introduces computing concepts using non-programming examples and artifacts.

CS0 An introductory college-level course on computing aimed at non-majors with little or no prior experience of programming.

CS1 An introductory college-level computer science course, typically one semester long, that focuses on variables, loops, functions, and other basic mechanics.

CS2 A second college-level computer science course that typically introduces basic data structures such as stacks, queues, and dictionaries.

Deficit model The idea that some groups are under represented in computing (or some other field) because their members lack some attribute or quality.

Deliberate practice The act of observing performance of a task while doing it in order to improve ability.

Demonstration lesson A staged lesson in which one teacher presents to actual learners while other teachers observe in order to learn new teaching techniques.

Diagnostic power The degree to which a wrong answer to a question or exercise tells the teacher what misconceptions a particular learner has.

Direct instruction A teaching method centered around meticulous curriculum design delivered through prescribed script.

Dunning-Kruger effect The tendency of people who only know a little about a subject to incorrectly estimate their understanding of it.

Educational psychology The study of how people learn. See also **instructional design**.

Ego depletion The impairment of self control that occurs when it is exercised intensively or for long periods.

Elevator pitch A short description of an idea, project, product, or person that can be delivered and understood in just a few seconds.

End-user programmer Someone who does not consider themselves a programmer, but who nevertheless writes and debugs software, such as an artist creating complex macros for a drawing tool.

End-user teacher By analogy with **end-user programmer**, someone who is teaching frequently, but whose primary occupation is not teaching, who has little or no background in pedagogy, and who may work outside institutional classrooms.

Expert Someone who can diagnose and handle unusual situations, knows when the usual rules do not apply, and tends to recognize solutions rather than reasoning to them. See also **competent practitioner** and **novice**.

Expert blind spot The inability of experts to empathize with novices who are encountering concepts or practices for the first time.

Expertise reversal effect The way in which instruction that is effective for novices becomes ineffective for competent practitioners or experts.

Externalized cognition The use of graphical, physical, or verbal aids to augment thinking.

Extraneous load Any **cognitive load** that distracts from learning.

Extrinsic motivation Being driven by external rewards such as payment or fear of punishment. See also **intrinsic motivation**.

Faded example A series of examples in which a steadily increasing number of key steps are blanked out. See also **scaffolding**.

False beginner Someone who has studied a language before but is learning it again. False beginners start at the same point as true beginners (i.e., a pre-test will show the same proficiency) but can move much more quickly.

Far transfer The **transfer of learning** between widely-separated domains, e.g., improvement in math skills as a result of playing chess.

Fink's Taxonomy A six-part non-hierarchical classification of understanding first proposed in [Fink2013] whose categories are *foundational knowledge, application, integration, human dimension, caring,* and *learning how to learn.* See also **Bloom's Taxonomy**.

Fixed mindset The belief that an ability is innate, and that failure is due to a lack of some necessary attribute. See also **growth mindset**.

Flipped classroom One in which learners watch recorded lessons on their own time, while class time is used to work through problem sets and answer questions.

Flow The feeling of being fully immersed in an activity; frequently associated with high productivity.

Fluid representation The ability to move quickly between different models of a problem.

Formative assessment Assessment that takes place during a lesson in order to give both the learner and the teacher feedback on actual understanding. See also **summative assessment**.

Free-range learner Someone learning outside an institutional classrooms with required homework and mandated curriculum. (Those who use the term occasionally refer to learners in classrooms as "battery-farmed learners," but we don't because that would be rude.)

Fuzz testing A software testing technique based on generating and submitting random data.

General pedagogical knowledge A person's understanding of the general principles of teaching. See also **content knowledge** and **pedagogical content knowledge**.

Germane load The **cognitive load** required to link new information to old.

Governance board A board whose primary responsibility is to hire, monitor, and if need be, fire the director.

Growth mindset The belief that ability comes with practice. See also **fixed mindset**.

Guided notes Teacher-prepared notes that cue learners to respond to key information in a lecture or discussion.

Hashing Generating a condensed pseudo-random digital key from data; any specific input produces the same output, but different inputs are highly likely to produce different outputs.

Hypercorrection effect The more strongly someone believed that their answer on a test was right, the more likely they are not to repeat the error once they discover that in fact they were wrong.

Implementation science The study of how to translate research findings to everyday clinical practice.

Impostor syndrome A feeling of insecurity about one's accomplishments that manifests as a fear of being exposed as a fraud.

Inclusivity Working actively to include people with diverse backgrounds and needs.

Inquiry-based learning The practice of allowing learners to ask their own questions, set their own goals, and find their own path through a subject.

Instructional design The craft of creating and evaluating specific lessons for specific audiences. See also **educational psychology**.

Intrinsic load The **cognitive load** required to absorb new information.

Intrinsic motivation Being driven by enjoyment of a task or the satisfaction of doing it for its own sake. See also **extrinsic motivation**.

Intuition The ability to understand something immediately, without any apparent need for conscious reasoning.

Jugyokenkyu Literally "lesson study," a set of practices that includes having teachers routinely observe one another and discuss lessons to share knowledge and improve skills.

Learned helplessness A situation in which people who are repeatedly subjected to negative feedback that they have no way to escape learn not to even try to escape when they could.

Learner persona A brief description of a typical target learner for a lesson that includes their general background, what they already know, what they want to do, how the lesson will help them, and any special needs they might have.

Learning Management System (LMS): An application for tracking course enrollment, exercise submissions, grades, and other bureaucratic aspects of formal classroom learning.

Learning objective What a lesson is trying to achieve.

Learning outcome What a lesson actually achieves.

Legitimate peripheral participation Newcomers' participation in simple, low-risk tasks that a **community of practice** recognizes as valid contributions.

Live coding The act of teaching programming by writing software in front of learners as the lesson progresses.

Long-term memory The part of memory that stores information for long periods of time. Long-term memory is very large, but slow. See also **short-term memory**.

Manual Reference material intended to help someone who already understands a subject fill in (or remind themselves of) details.

Marketing The craft of seeing things from other people's perspective, understanding their wants and needs, and finding ways to meet them

Massive Open Online Course (MOOC) An online course designed for massive enrollment and asynchronous study, typically using recorded videos and automated grading.

Mental model A simplified representation of the key elements and relationships of some problem domain that is good enough to support problem solving.

Metacognition Thinking about thinking.

Minimal manual An approach to training that breaks every task into single-page instructions that also explain how to diagnose and correct common errors.

Minute cards A feedback technique in which learners spend a minute writing one positive thing about a lesson (e.g., one thing they've learned) and one negative thing (e.g., a question that still hasn't been answered).

Near transfer The **transfer of learning** between closely-related domains, e.g., improvement in understanding of decimals as a result of doing exercises with fractions.

Notional machine A general, simplified model of how a particular family of programs executes.

Novice Someone who has not yet built a usable mental model of a domain. See also **competent practitioner** and **expert**.

Objects first An approach to teaching programming in which objects and classes are introduced early.

Pair programming A software development practice in which two programmers share one computer. One programmer (the driver) does the typing, while the other (the navigator) offers comments and suggestions in real time. Pair programming is often used as a teaching practice in programming classes.

Parsons Problem An assessment technique developed by Dale Parsons and others in which learners rearrange given material to construct a correct answer to a question.

Passive learning An approach to instruction in which learners read, listen, or watch without immediately using new knowledge. Passive learning is less effective than **active learning**.

Passive teaching An approach to instruction in which the teacher does not adjust pace or examples, or otherwise act on feedback from learners, during the lesson. See also **active teaching**.

Pedagogical content knowledge (PCK) The understanding of how to teach a particular subject, i.e., the best order in which to introduce topics and what examples to use. See also **content knowledge** and **general pedagogical knowledge**.

Peer instruction A teaching method in which an teacher poses a question and then learners commit to a first answer, discuss answers with their peers, and commit to a (revised) answer.

Persistent memory See **long-term memory**.

Personalized learning Automatically tailoring lessons to meet the needs of individual learners.

Plausible distractor A wrong or less-than-best answer to a multiple-choice question that looks like it could be right. See also **diagnostic power**.

Positioning What sets one brand apart from other, similar brands.

Preparatory privilege The advantage of coming from a background that provides more preparation for a particular learning task than others.

Productive failure A situation in which learners are deliberately given problems that can't be solved with the knowledge they have and must go out and acquire new information in order to make progress. See also **Zone of Proximal Development**.

Pull request A set of proposed changes to a GitHub repository that can be reviewed, updated, and eventually merged.

Read-cover-retrieve A study practice in which the learner covers up key facts or terms during a first pass through material, then checks their recall on a second pass.

Reflective practice See **deliberate practice**.

Reusability Paradox Holds that the more reusable part of a lesson is, the less pedagogically effective it is.

Scaffolding Extra material provided to early-stage learners to help them solve problems.

Search engine optimization Increasing the quantity and quality of website traffic by making pages more easily found by, or seem more important to, search engines.

Service board A board whose members take on working roles in the organization.

Short-term memory The part of memory that briefly stores information that can be directly accessed by consciousness.

Situated learning A model of learning that focuses on people's transition from being newcomers to be accepted members of a **community of practice**.

Split-attention effect The decrease in learning that occurs when learners must divide their attention between multiple concurrent presentations of the same information (e.g., captions and a voiceover).

Stereotype threat A situation in which people feel that they are at risk of being held to stereotypes of their social group.

Subgoal labeling Giving names to the steps in a step-by-step description of a problem-solving process.

Summative assessment Assessment that takes place at the end of a lesson to tell whether the desired learning has taken place.

Tangible artifact Something a learner can work on whose state gives feedback about the learner's progress and helps the learner diagnose mistakes.

Teaching to the test Any method of "education" that focuses on preparing students to pass standardized tests, rather than on actual learning.

Test-driven development A software development practice in which programmers write tests first in order to give themselves concrete goals and clarify their understanding of what "done" looks like.

Think-pair-share A collaboration method in which each person thinks individually about a question or problem, then pairs with a partner to pool ideas, and then have one person from each pair present to the whole group.

Transfer of learning Applying knowledge learned in one context to problems in another context. See also **near transfer** and **far transfer**.

Transfer-appropriate processing The improvement in recall that occurs when practice uses activities similar to those used in testing.

Tutorial A lesson intended to help someone improve their general understanding of a subject.

Twitch coding Having a group of people decide moment by moment or line by line what to add to a program next.

Working memory See **short-term memory**.

Zone of Proximal Development (ZPD) Includes the set of problems that people cannot yet solve on their own but are able to solve with assistance from a more experienced mentor. See also **productive failure**.

E Meetings, Meetings, Meetings

Most people are really bad at meetings: they don't make an agenda, they don't take minutes, they waffle on or wander off into irrelevancies, they say something banal or repeat what others have said just so that they'll have said something, and they hold side conversations (which pretty much guarantees that the meeting will be a waste of time). Knowing how to run a meeting efficiently is a core skill for anyone who wants to get things done; knowing how to take part in someone else's meeting is just as important (though it gets far less attention—as a colleague once said, everyone offers leadership training, but nobody offers followership training).

The most important rules for making meetings efficient are not secret, but are rarely followed:

Decide if there actually needs to be a meeting. If the only purpose is to share information, send a brief email instead. Remember, you can read faster than anyone can speak: if someone has facts for the rest of the team to absorb, the most polite way to communicate them is to write them up.

Write an agenda. If nobody cares enough about the meeting to write a point-form list of what's to be discussed, the meeting probably doesn't need to happen.

Include timings in the agenda. Agendas can also help you prevent early items stealing time from later ones if you include the time to be spent on each item in the agenda. Your first estimates with any new group will be wildly optimistic, so revise them upward for subsequent meetings. However, you shouldn't plan a second or third meeting because the first one ran over-time: instead, try to figure out why you're running over and fix the underlying problem.

Prioritize. Every meeting is a micro-project, so work should be prioritized in the same way that it is for other projects: things that will have high impact but take little time should be done first, and things that will take lots of time but have little impact should be skipped.

Make one person responsible for keeping things moving. One person should be tasked with keeping items to time, chiding people who are checking email or having side conversations, asking people who are talking too much to get to the point, and inviting people who aren't talking to express an opinion. This person should *not* do all the talking; in fact, whoever is in charge will talk less in a well-run meeting than most other participants.

Require politeness. No one gets to be rude, no one gets to ramble, and if someone goes off topic, it is both the moderator's right and responsibility to say, "Let's discuss that elsewhere."

No interruptions. Participants should raise a finger or put up a sticky note if they want to speak next. If the person speaking doesn't notice them, the moderator should.

No technology unless it's required for accessibility reasons. Insist that everyone put their phones, tablets, and laptops into politeness mode (i.e., close them).

Record minutes. Someone other than the moderator should take point-form notes about the most important pieces of information that were shared, every decision that was made, and every task that was assigned to someone.

Take notes. While other people are talking, participants should take notes of questions they want to ask or points they want to make. (You'll be surprised how smart it makes you look when it's your turn to speak.)

End early. If your meeting is scheduled for 10:00-11:00, you should aim to end at 10:50 to give people time to visit the bathroom on their way to where they need to go next.

As soon as the meeting is over, email the minutes to everyone or post them on the web:

People who weren't at the meeting can keep track of what's going on. A web page or email message is a much more efficient way to catch up than asking a team mate what you missed.

Everyone can check what was actually said or promised. More than once, I have looked over the minutes of a meeting I was in and thought, "Did I say that?" or, "Wait a minute, I didn't promise to have it ready then!" Accidentally or not, people will often remember things differently; writing it down gives team members a chance to correct mistakes, which can save a lot of anguish later on.

People can be held accountable at subsequent meetings. There's no point making lists of questions and action items if you don't follow up on them later. If you're using some kind of issue tracking system, create an issue for each new question or task right after the meeting and update those that are being carried forward, then start each meeting by going through a list of those issues.

[Brow2007; Broo2016; Roge2018] have lots of advice on running meetings. In my experience, an hour of training on how to be a moderator is one of the best investments you will ever make.

Sticky Notes and Interruption Bingo

Some people are so used to the sound of their own voice that they will insist on talking half the time no matter how many other people are in the room. To combat this, give everyone three sticky notes at the start of the meeting. Every time they speak, they have to take down one sticky note. When they're out of notes, they aren't allowed to speak until everyone has used at least one, at which point everyone gets all of their sticky notes back. This ensures that nobody talks more than three times as often as the quietest person in the meeting, and completely changes the dynamics of most groups: people who have given up trying to be heard because they always get trampled suddenly have space to contribute, and the overly-frequent speakers quickly realize just how unfair they have been[1].

Another technique is interruption bingo. Draw a grid and label the rows and columns with participants' names. Add a tally mark to the appropriate cell each time someone interrupts someone else, and take a moment to share the results halfway through the meeting. In most cases, you will see that one or two people are doing all of the interrupting, often without being aware of it. That alone is often enough to get them to throttle back. (Note that this technique is intended to manage interruptions, not speaking time: it may be appropriate for people with more knowledge of a subject to speak about it more often in a meeting, but it is never appropriate to repeatedly cut people off.)

E.1 MARTHA'S RULES

Organizations all over the world run their meetings according to Robert's Rules of Order[2], but they are far more formal than most small projects require. A lightweight alternative known as "Martha's Rules" may be much better for consensus-based decision making [Mina1986]:

1. Before each meeting, anyone who wishes may sponsor a proposal by sharing it with the group. Proposals must be filed at least 24 hours before a meeting in order to be considered at that meeting and must include:
   - a one-line summary;
   - the full text of the proposal;
   - any required background information;
   - pros and cons; and
   - possible alternatives
   Proposals should be at most two pages long.
2. A quorum is established in a meeting if half or more of voting members are present.
3. Once a person has sponsored a proposal, they are responsible for it. The group may not discuss or vote on the issue unless the sponsor or their delegate is present. The sponsor is also responsible for presenting the item to the group.
4. After the sponsor presents the proposal, a preliminary vote is cast for the proposal prior to any discussion:
   - Who likes the proposal?
   - Who can live with the proposal?
   - Who is uncomfortable with the proposal?

---

[1] I certainly did when this was done to me...

[2] https://en.wikipedia.org/wiki/Robert%27s_Rules_of_Order

Preliminary votes can be done thumb up, thumb sideways, or thumb down (in person) or by typing +1, 0, or -1 into online chat (in virtual meetings).

5. If all or most of the group likes or can live with the proposal, it is immediately moved to a formal vote with no further discussion.
6. If most of the group is uncomfortable with the proposal, it is postponed for further rework by the sponsor.
7. If some members are uncomfortable they can briefly state their objections. A timer is then set for a brief discussion moderated by the facilitator. After ten minutes or when no one has anything further to add (whichever comes first), the facilitator calls for a yes-or-no vote on the question: "Should we implement this decision over the stated objections?" If a majority votes "yes" the proposal is implemented. Otherwise, the proposal is returned to the sponsor for further work.

E.2 ONLINE MEETINGS

Chelsea Troy's discussion[3] of why online meetings are often frustrating and unproductive makes an important point: in most online meetings, the first person to speak during a pause gets the floor. The result? "If you have something you want to say, you have to stop listening to the person currently speaking and instead focus on when they're gonna pause or finish so you can leap into that nanosecond of silence and be the first to utter something. The format… encourages participants who want to contribute to say more and listen less."

The solution is to run a text chat beside the video conference where people can signal that they want to speak, The moderator then selects people from the waiting list. If the meeting is large or argumentative, have everyone mute themselves and only allow the moderator to unmute people.

E.3 THE POST MORTEM

Every project should end with a post mortem in which participants reflect on what they just accomplished and what they could do better next time. Its aim is *not* to point the finger of shame at individuals, although if that has to happen, the post mortem is the best place for it.

A post mortem is run like any other meeting with a few additional guidelines [Derb2006]:

Get a moderator who wasn't part of the project and doesn't have a stake in it.

Set aside an hour, and only an hour. In my experience, nothing useful is said in the first ten minutes of anyone's first post mortem, since people are naturally a bit shy about praising or damning their own work. Equally, nothing useful is said after the first hour: if you're still talking, it's probably because one or two people have things they want to get off their chests rather than suggestions for improvements.

[3]https://chelseatroy.com/2018/03/29/why-do-remote-meetings-suck-so-much/

Require attendance. Everyone who was part of the project ought to be in the room for the post mortem. This is more important than you might think: the people who have the most to learn from the post mortem are often least likely to show up if the meeting is optional.

Make two lists. When I'm moderating, I put the headings "Do Again" and "Do Differently" on the board, then ask each person to give me one item for each list in turn without repeating anything that has already been said.

Comment on actions rather than individuals. By the time the project is done, some people may no longer be friends. Don't let this sidetrack the meeting: if someone has a specific complaint about another member of the team, require them to criticize a particular event or decision. "They had a bad attitude" does *not* help anyone improve.

Prioritize the recommendations. Once everyone's thoughts are out in the open, sort them according to which are most important to keep doing and which are most important to change. You will probably only be able to tackle one or two from each list in your next project, but if you do that every time, your life will quickly get better.

F Checklists and Templates

[Gawa2007] popularized the idea that using checklists can save lives, and more recent studies have generally supported their effectiveness [Avel2013; Urba2014; Rams2019]. We find them useful, particularly when bringing new teachers onto a team; the ones given below can be used as starting points for developing your own.

F.1 TEACHING EVALUATION

This rubric is designed to assess people teaching for 5–10 minutes with slides, live coding, or a mix of both. Rate each item as "Yes," "Iffy," "No," or "Not Applicable."

Opening	Exists (use N/A for other responses if not)
	Good length (10–30 seconds)
	Introduces self
	Introduces topics to be covered
	Describes prerequisites
Content	Clear goal/narrative arc
	Inclusive language
	Authentic tasks/examples
	Teaches best practices/uses idiomatic code
	Steers a path between the Scylla of jargon and the Charybdis of over-simplification
Delivery	Clear, intelligible voice (use "Iffy" or "No" for strong accent)
	Rhythm: not too fast or too slow, no long pauses or self-interruption, not obviously reading from a script
	Self-assured: does not stray into the icky tarpit of uncertainty or the dungheap of condescension
Slides	Exist (use N/A for other responses if not)
	Slides and speech complement one another (dual coding)
	Readable fonts and colors/no overwhelming slabs of text
	Frequent change on screen (something every 30 seconds)
	Good use of graphics
Live Coding	Used (use N/A for other responses if not)
	Code and speech complement one another
	Readable fonts and colors/right amount of code on the screen
	Proficient use of tools
	Highlights key features of code
	Dissects errors

Closing	Exists (use N/A for other responses if it doesn't) Good length (10–30 seconds) Summarizes key points Outlines next steps
Overall	Points clearly connected/logical flow Make the topic interesting (i.e., not boring) Knowledgeable

F.2 TEAMWORK EVALUATION

This rubric is designed to assess individual performance within a team. You can use it as a starting point for creating a rubric of your own. Rate each item as "Yes," "Iffy," "No," or "Not Applicable."

Communication	Listens attentively to others without interrupting Clarifies with others have said to ensure understanding Articulates ideas clearly and concisely Gives good reasons for ideas Wins support from others
Decision Making	Analyzes problems from different points of view Applies logic in solving problems Offers solutions based on facts rather than "gut feel" or intuition Solicits new ideas from others Generates new ideas Accepts change
Collaboration	Acknowledges issues that the team needs to confront and resolve Works toward solutions that are acceptable to all involved Shares credit for success with others Encourages participation among all participants Accepts criticism openly and non-defensively Cooperates with others
Self-Management	Monitors progress to ensure that goals are met Puts top priority on getting results Defines task priorities for work sessions Encourages others to express views even when they are contrary Stays focused on the task during meetings Uses meeting time efficiently Suggests ways to proceed during work sessions

F.3   EVENT SETUP

The checklists below are used before, during, and after events.

SCHEDULING THE EVENT

- Decide if it will be in person, online for one site, or online for several sites.
- Talk through expectations with the host and make sure everyone agrees who is covering travel costs.
- Determine who is allowed to take part: is the event open to all comers, restricted to members of one organization, or something in between?
- Arrange teachers.
- Arrange space, including breakout rooms if needed.
- Choose dates. If it is in person, book travel.
- Get names and email addresses of attendees from host.
- Make sure everyone is registered.

SETTING UP

- Set up a web page with details on the workshop, including date, location, and what participants need to bring.
- Check whether any attendees have special needs.
- If the workshop is online, test the video conferencing—twice.
- Make sure attendees will have network access.
- Create a page for sharing notes and exercise solutions (e.g., a Google Doc).
- Email attendees a welcome message with a link to the workshop page, background readings, a description of any setup they need to do, a list of what they need to bring, and a way to contact the host or teacher on the day.

AT THE START OF THE EVENT

- Remind everyone of the code of conduct.
- Take attendance and create a list of names to paste into the shared page.
- Distribute sticky notes.
- Make sure everyone can get online.
- Make sure everyone can access the shared page.
- Collect any relevant online account IDs.

AT THE END OF THE EVENT

- Update the attendance list.
- Collect feedback from participants.
- Make a copy of the shared page.

TRAVEL KIT

Here are a few things teachers take with them to workshops:

sticky notes	cough drops
comfortable shoes	a small notepad
a spare power adapter	a spare shirt
deodorant	video adapters
laptop stickers	their notes (printed or on a tablet)
a granola bar or some other snack	antacid (because road food)
business cards	spare glasses/contacts
a notebook and pen	a laser pointer
an insulated cup for tea/coffee	extra whiteboard markers
a toothbrush or mouthwash	wet wipes (because spills happen)

When traveling, many teachers also take running shoes, a bathing suit, a yoga mat, or whatever else they exercise in or with. Some also bring a portable WiFi hub in case the room's network isn't working and some USB drives with installers for the software learners need.

F.4   LESSON DESIGN

This section summarizes the backward design method developed independently by [Wigg2005; Bigg2011; Fink2013]. It lays out a step-by-step progression to help you think about the right things in the right order and provides spaced deliverables so you can re-scope or redirect effort without too many unpleasant surprises.

Everything from Step 2 onward goes into your final lesson, so there is no wasted effort; as described in Chapter 6, writing sample exercises early helps ensure that everything you ask your learners to do contributes to the lesson's goals and that everything they need to know is covered.

The steps are described in order of increasing detail, but the process itself is always iterative. You will frequently go back to revise earlier work as you learn something from your answers to later questions or realize that your initial plan isn't going to play out the way you first thought.

WHO IS THIS LESSON FOR?

Create some learner personas (Section 6.1) or (preferably) choose ones that you and your colleagues have drawn up for general use. Each persona should have:

1. the person's general background;
2. what they already know;
3. what they think they want to do; and
4. any special needs they have.

Deliverable: brief summaries of who you are trying to help.

WHAT'S THE BIG IDEA?

Write point-form answers to three or four of the questions below to help you figure out the scope of the lesson. You don't need to answer all of these questions, and you may pose and answer others if you think it's helpful, but you should always include a couple of answers to the first one. You may also create a concept map at this stage (Section 3.1).

- What problems will people learn how to solve?
- What concepts and techniques will people learn?
- What technologies, packages, or functions will people use?
- What terms or jargon will you define?
- What analogies will you use to explain concepts?
- What mistakes or misconceptions do you expect?
- What datasets will you use?

Deliverable: a rough scope for the lesson. Share this with a colleague—a little bit of feedback at this point can save hours of wasted effort later on.

WHAT WILL LEARNERS DO ALONG THE WAY?

Make the goals in Step 2 firmer by writing full descriptions of a couple of exercises that learners will be able to do toward the end of the lesson. Doing this is analogous to test-driven development[1]: rather than working forward from a (probably ambiguous) set of learning objectives, work backward from concrete examples of where you want your learners to end up. Doing this also helps uncover technical requirements that might otherwise not be found until uncomfortably late.

To complement the full exercise descriptions, write brief point-form descriptions of one or two exercises per lecture hour to show how quickly you expect learners to progress. Again, these serve as a good reality check on how much you're assuming and help uncover technical requirements. One way to create these "extra" exercises is to make a point-form list of the skills needed to solve the major exercises and create an exercise that targets each.

Deliverable: 1–2 fully explained exercises that use the skills people are to learn, plus half a dozen point-form exercise outlines. Include complete solutions so that you can make sure the software you want learners to use actually works.

HOW ARE CONCEPTS CONNECTED?

Put the exercises you have created in a logical order and then derive a point-form lesson outline from them. The outline should have 3–4 bullet points for each hour with

---

[1] https://en.wikipedia.org/wiki/Test-driven_development

a formative assessment of some kind for each. It is common to change assessments in this stage so that they can build on each other.

Deliverable: a lesson outline. You are likely to discover things you forgot to list earlier during this stage, so don't be surprised if you have to double back a few times.

LESSON OVERVIEW

You can now write a lesson overview with:

- a one-paragraph description (i.e., a sales pitch to learners);
- half a dozen learning objectives; and
- a summary of prerequisites.

Doing this earlier often wastes effort, since material is usually added, cut, or moved around in earlier steps.

Deliverable: course description, learning objectives, and prerequisites.

F.5   PRE-ASSESSMENT QUESTIONNAIRE

This questionnaire helps teachers gauge the prior programming knowledge of participants in an introductory JavaScript workshop. The questions and answers are concrete, and the whole thing is short so that respondents won't find it intimidating.

1. Which of these best describes your experience with programming in general?
    - I have none.
    - I have written a few lines now and again.
    - I have written programs for my own use that are a couple of pages long.
    - I have written and maintained larger pieces of software.

2. Which of these best describes your experience with programming in JavaScript?
    - I have none.
    - I have written a few lines now and again.
    - I have written programs for my own use that are a couple of pages long.
    - I have written and maintained larger pieces of software.

3. Which of these best describes how easily you could write a program in any language to find the largest number in a list?
    - I wouldn't know where to start.
    - I could struggle through by trial and error with a lot of web searches.
    - I could do it quickly with little or no use of external help.

4. Which of these best describes how easily you could write a JavaScript program to find and capitalize all of the titles in a web page?
   - I wouldn't know where to start.
   - I could struggle through by trial and error with a lot of web searches.
   - I could do it quickly with little or no use of external help.

5. What do you want to know or be able to do after this class that you don't know or can't do right now?

G Example Concept Maps

These concept maps were created by Amy Hodge of Stanford University, and are re-used with permission.

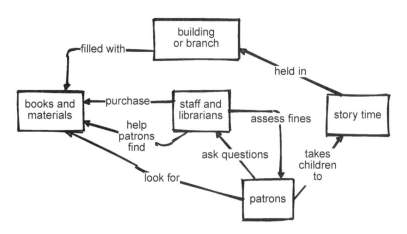

Figure G.1: Library patron concept map

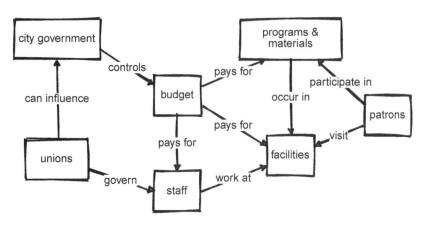

Figure G.2: Library director concept map

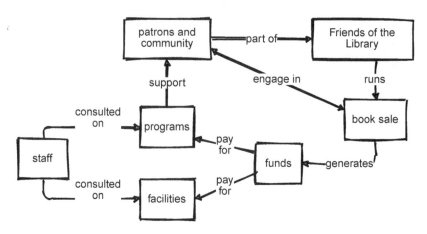

Figure G.3: Library friends concept map

H Chunking Exercise Solution

See the last exercise in Chapter 3 for the unchunked representation of these symbols.

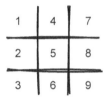

Figure H.1: Chunked representation

Index